DARK WINTER

How the Sun Is Causing a 30-Year Cold Spell

DARK WINTER

How the Sun Is Causing a 30-Year Cold Spell

by John L. Casey

Humanix Books
www.humanixbooks.com
Boca Raton, FL, USA

Dark Winter: How the Sun Is Causing a 30-Year Cold Spell

Humanix Books
P.O. Box 20989
West Palm Beach, FL 33416
USA
www.humanixbooks.com
email: info@humanixbooks.com

Printed in the United States of America and the United Kingdom.

ISBN (Hardcover) 978-1-63006-035-0
ISBN (E-book) 978-1-63006-023-7

Library of Congress Control Number 2014939082

Contents

Foreword

THE MATTER OF CLIMATE CHANGE has been a perplexing problem for decades. Do human behaviors and social dynamics follow environmental changes, or is it the contrary — that human activity is the leader of the relationship between humans and the environment? The "modern," popular belief supported by the media, politicians, and some scientists is that humans control the environment. Is it true?

Of course not! The Earth's biosphere development is a result of about three billion years of changes in the environment of our planet. The final product of this evolution up to this moment is the human species. Thus, humans are a result of these environmental changes, not the cause of them.

One of the key features of climate change is relatively well-expressed cycles. These cycles are caused mainly by celestial factors: variations of the Earth's motion, solar activity, and so on. The Milankovitch theory describes the cyclic variations of Earth orbital and rotational motion parameters in cycles up to 120,000 years. But while the theory explains satisfactorily such important climatic events as the "great ice epochs" and other large climate cycles, it is impossible to use as a tool for understanding the shorter climate cycles, such as those lasting from 2 to 10,000 years.

The most important of these are the so-called "cycles of the little ice epochs" (2,200 to 2,400 years), as well as climate oscillations of about 200 years. The little ice epoch cycle was first detected by geologists George Denton and Wibjörn Karlén at the end of the 1960s by analysis of the dynamics of the edges of high mountain ice fields. In addition, this 2,200- to 2,400-year cycle has been found in a large number of indirect climate parameters: the world ocean level changes, the level of isolated basins like the Caspian Sea, and the growth of stalactites and stalagmites, as well as in many tree samples. The 2,200- to 2,400-year cycle is a major feature of the global climate. It corresponds to temperature variations of 2°C to 2.5°C. The little ice ages are the starting epochs of this cycle. The last one occurred in the fifteenth to seventeenth century.

The Bicentennial Cycle, with a duration of about 200 years, comes with a global temperature change of 0.5°C to 1°C and is also well-expressed in global climates, but mainly in the inner parts of continents — for example, Eurasia and North America. There is now much evidence that the 2,200- to 2,400-year cycle and the 200-year cycle, as well as almost all other relatively short climatic cycles, are caused by processes on the sun.

There have been many scientists going back to the 1950s and 1960s who have established the relationship between solar climate and 200-year cycles (e.g., many experts before 1963 in the former USSR and Czechoslovakia; many authors before 1963; Paul Damon of the United States in 1968; and A. Bonov of Bulgaria in 1968). The relationship between the "Maunder-type" minimums (or "prolonged sunspot minimums") of the solar 2,200- to 2,400-year cycle and the little ice epochs was well-established and described at the beginning of the 1990s in the papers of Damon and Charles Sonett in the United States and Valentin Dergachev and Vladimir Chistyakov in Russia.

However, the role of the sun in climate change up to this moment has been studied and debated by only a relatively small group of scientists — mainly astrophysicists, geophysicists, and paleoclimatologists. It is a "forbidden theme" for researchers connected to the World Meteorological Organization (WMO) and the United Nations Intergovernmental Panel on Climate Change

(IPCC), as well as for the media. As a result, the influence of the sun on climate change is almost unknown for the great majority of people not only in the United States but in the rest of the world. An additional problem is a lack of awareness in education in the fields of physics, astronomy, geology, and other natural and "space-oriented" disciplines. This learning deficit is seen in our middle schools, colleges, and most universities, except in the most advanced and specialized institutions.

In this regard, John L. Casey's *Dark Winter* is a very useful book. It is written for the layman and thus is not a scientific textbook. Its style speaks very well to educated but nonscientific readers. It is, therefore, ideal for students in high schools, colleges, and universities; educators; economists; agro- and hydro-engineers; developers of energy strategies; politicians; businessmen; and any others who want to understand the fuller picture of climate change. It is especially recommended to journalists who are interested in reporting on the field of climate change.

Casey's primary argument is that a longtime downward tendency in solar activity dynamics has already started, and it will cause a corresponding climate cooling similar to the Dalton Minimum. I agree, though the real cooling effect, in my opinion, would more probably be 0.7°C to 0.8°C. Others have also predicted this coming cold era as forecast by Casey, including M. V. Fyodorov et al. (*Solar Physics,* 1996), Drs. Boris Komitov and B. P. Bonev (*Astrophysical Journal,* 2001), and M. G. Ogurtsov (*Solar Physics,* 2005).

Casey presents a thought-provoking analysis of the social conditions in Europe and the United States at the beginning of nineteenth century and their relation to the Dalton Minimum and corresponding climate situation. The effect of Indonesia's Mount Tambora eruption in 1815 is also discussed. The book brings up, if only briefly, an interesting question: Is there a relationship between solar activity and Earth tectonics? According to a 2006 study by the Russian geologists Rogosin and Shestopalov, the overall seismic activity of the Earth essentially increases during these long-term solar minimums, just as Casey describes. In this case, I believe that Casey is correct in that there is an increasing probability of major earthquakes and volcanic eruptions, especially in the Pacific region, during this next solar minimum.

The economic and social effects of the forthcoming solar minimum are well discussed in the book. It should be emphasized that major demographic and social events in the history of the Old World (i.e., Europe, Asia, and Africa) during the last 5,000 years coincide strongly with solar activity and corresponding climatic changes. For example, the Renaissance and little ice age epochs coincide with the Maunder Minimums. These periods saw the largest gains in human progress, but they also saw significant food and energy shortages. I have also found this correlation in my own research.

The problems brought on by this next period of climate change, which Casey ably discusses in *Dark Winter*, are more potentially troublesome than ever before. We are once again facing food and energy shortages, but this time with a population upwards of seven billion people. This also comes at a time when our technical infrastructure is more susceptible to environmental, climate, and tectonic disruptions.

For these reasons, I recommend the book *Dark Winter* as being both timely and necessary.

Dr. Boris Komitov, Bulgarian Academy of Sciences

Preface

DARK *WINTER* COMES AT a unique time in the ongoing debate about climate change in this country and around the world. As this book goes to press, we are emerging from yet another record-setting cold winter in the northern hemisphere. Further, confirmation has been gleaned from measured climate parameters of an ongoing transition from the past, naturally caused, globally warm period to a returning, solar-induced, cold climate epoch. This book chronicles the history and science behind this irrefutable trend, drawing on accurate, proven, and highly reliable climate models based on the Sun's behavior — models that have been simply ignored by those who drive climate politics in the United Nations and US government.

Regardless of the overwhelming evidence to the contrary, we continue to see UN and US governmental policy based on the now thoroughly discredited greenhouse gas theory and the insignificant role mankind's industrial CO_2 emissions play in the atmosphere. In what can only be classified as a nationwide fit of cognitive dissonance in this country, many of our leaders, including the president and secretary of state and members of the media, have resorted to reinforcing the now disproved myth of man-made global warming with outlandish claims and outright lies about the state of the Earth's climate and where it's headed. In a predictable move to discredit those who rely on the facts — not the politics — of climate change, these same leaders have taken to personal attacks and name calling, labeling those who reject the politically correct version of climate science as "members of the Flat Earth Society" and, even more insidiously, "deniers," attempting to associate climate truth seekers with those who dispute the reality of the Holocaust.

This effort perhaps reached its zenith with President Barack Obama's Georgetown University address on June 25, 2013, when he announced his "Climate Action Plan." It signaled the latest salvo in what has become a juggernaut of political arm-twisting and media-distributed propaganda regarding alleged man-made global warming (aka "climate change") to squeeze ever more political power and money out of American taxpayers in order to pursue a predetermined agenda.

But the president, his supporters in the media, and even the United Nations Intergovernmental Panel on Climate Change (IPCC) know full well that there are a few immutable facts that are about to end the man-made climate change charade:

- **Climate change is simply not important to a growing number of voters.** According to recent Pew and WSJ/NBC polls, people rate climate change as dead last on their list of concerns. They just don't see it as a threat. Some of the reasons include: 1) it doesn't literally have an impact on their daily lives; 2) they have much more immediate issues they have to deal with; 3) they don't accept the whole "man-made climate change" myth anyway; and 4) like many important political issues, they feel they have no control over the outcome, regardless of their degree of passion and action.

- **Major changes taking place in the climate have already pulled the rug out from under the idea that mankind controls the climate.** Perhaps the most glaring example is that there has been no global warming for seventeen years now. That means that while we have been browbeaten over mankind's CO_2 emissions and the threat of global warming, the globe hasn't even been warming to begin with! Further, the actual record of global temperature trends shows the oceans and the atmosphere have actually been cooling for most of the last 11 years. The end result: global warming has ended and a new cold climate has begun! This fundamental, inescapable conclusion flies in the face of assertions by the president and IPCC when they say global warming is "accelerating," or that they are now even more convinced (with 95 percent certainty) that mankind is causing global warming.

- **A growing group of "global cooling" scientists is issuing a serious climate alert, and the people are starting to listen to them.** This group of mostly international scientists, to which I belong, is concerned that we are wasting valuable time in a pointless debate when the precious time to prepare for a globally destructive cold epoch dwindles. We now have members of the Russian Academy of Sciences openly saying a new "Little Ice Age" will start this year!
- **A possible shift in the political balance of power in the United States could end use of the flawed greenhouse gas theory and man's insignificant output of industrial CO_2 as a political tool.** As the truth about climate change becomes well known, a reordering of US government political power may derail any future plans for imposing new regulations on businesses and individuals through Environmental Protection Agency (EPA) edicts or climate-related legislation. If this occurs, climate change will no longer be used as a tool to transform America into a European-style, socialist form of government, and efforts to engage in a global redistribution of wealth through the fraudulent tool of man-made climate change will be effectively over, perhaps for good.

IPCC and US government devotion to the greenhouse gas theory of climate change is particularly troubling because the theory has been shown to be an abject failure of historic proportions.

Sadly, through the "Climategate" e-mails and many other disclosures of scientific misconduct, we now know the "settled" science was never about the science in the first place. What is truly settled in the climate debate is that, after almost 25 years of effort and tens of billions of dollars spent trying to make the greenhouse gas theory work, this "sow's ear" cannot be remade into a "silk purse." The many models developed to explain past climate behavior and predict future global temperature trends are significantly in error and produce results that do not represent the real world by a wide margin. They are so unreliable as to make the entire process of greenhouse gas climate assessment unusable for policy purposes.

Dark Winter was written to fill that void. It is an accurate, unbiased assessment of where the climate is going based on the most reliable, proven models for climate change — those based on solar activity. It is my hope that scientists, researchers, policymakers, journalists, and concerned citizens will read this book, absorb its findings, and act on its conclusions before it's too late.

John L. Casey
Editor, Global Climate Status Report
President, Space and Science Research Corporation

This Book Is Dedicated To . . .

. . . my late father, Edward A. Casey, from whom I learned that integrity is the most important quality one can possess, and my mother, Juanita L. Casey, from whom I learned the value of hard work and perseverance.

. . . my family, who believed in me throughout this book's long development, including my daughters, Shannon and Erin; Shannon's husband, Dave; and especially my wife, Alicia.

. . . the American people and people around the world who have been kept in the dark too long about what is really happening with our planet.

Introduction

"What do we live for if not to make life less difficult for each other?"

— *George Elliot*

THIS BOOK MAY BE the most important one I will ever write. In terms of its impact on your life, that of your friends and family and the lives of your descendants, it may also be one of the most important you will ever read.

For in this book, I will do nothing less than tell you what the near future holds for our planet from a climatic and geological standpoint, and how that, in turn, will provide a picture of your world — in effect, your future for the challenging decades ahead.

Here, I will present to you the primary theme of this book, so that there should be no doubt or confusion as to its message:

A historic reduction in the energy output of the Sun has begun. The most likely outcome from this "solar hibernation" will be widespread global loss of life, and social, economic, and political disruption. You must begin to prepare for this life-altering event now!

In this book, you will be given the full story behind the momentous arrival of this solar hibernation. You will read about the very day and time when I first discovered the cycles of the Sun that are causing this solar event that is going to change the world for all humans. You will learn that many other scientists have discovered that this significant climate change event is coming. This vital information should have been passed on to you many years ago.

You will be told of the ill effects we should expect for our planet in the next decades and read about what happened the last time this solar phenomenon occurred. We will examine the multiple disasters we have already started to face, including global food shortages and catastrophic earthquakes.

Most importantly, I will provide you with the mountain of evidence that shows that global warming has ended, and a new and potentially dangerous cold climate has begun. After that, you will be presented with some of my thoughts on what, if anything, we can do to prepare for the inevitable cold and difficult years to come. The book will close with a long-range view of Earth's future climate changes, predestined by repeating cycles of the Sun.

Those of you who have always looked skeptically at the overzealous efforts of politicians and the United Nations to dominate our lives or simply do not believe that mankind can control the Earth's climate may find this text easy to accept. Until I began my own, in-depth study of the Sun's cycles of behavior, I was generally in agreement with the consensus, if a bit suspiciously. This was the result of getting only one side — the wrong side — of the climate change story. After my research was done, all of that changed . . . so much so that I began a personal crusade to tell the world the true story about our climate's future — one that most have not heard, yet one that will affect us all.

In April and May of 2007, I became the first researcher in the United States to notify the White House, Congress, the mainstream media, all state governments, and the public of the dangers of the new cold era. At that time, I also made several major, specific predictions about the timing and character of this next climate era. Those predictions involved the following issues:

1. The end of global warming
2. Solar hibernation — a historic reduction in the energy output of the Sun that occurs every 206 years
3. A long-term drop in the Earth's average temperature
4. The advent of the next climate change, predicting 20 to 30 years of deep and dangerously cold weather

My predictions did not stop there. In May 2010, I stated that we would soon experience earthquakes and volcanic eruptions of historic scale. Only ten months later, on March 11, 2011, a catastrophic magnitude 9.0 earthquake struck Japan — the fifth largest in the past hundred years. The quake and resulting tsunami caused thousands to lose their lives and devastated much of the northeastern coast of Japan, wiping out many communities and threatening many nuclear reactors with meltdown. What is central to this predictive success is the belief that, to best understand Earth's climate changes and much of its geophysical processes, we only need to study the natural cycles of the Sun — it's that simple.

This book is also written as a personal story of my discovery of the Sun's cycles of activity that control our climate changes. I will discuss with you the Relational Cycle (RC) theory, which came from my research into these crucial cycles. This new, powerful theory and model for climate change now gives us another valuable tool for understanding nothing less than the schedule and amplitude of future climate changes. This book also relies heavily on many other respected US and international sources and scientists for opinions on global climate and geophysical processes. *Dark Winter,* as you will see, is far from being just about one researcher's theory.

First and foremost, what you are about to read regarding climate change is unvarnished, with no punches pulled; it will convey information you probably have not read in newspapers, seen on TV, or heard from our national leadership, except for a gallant few sources. Important findings and conclusions are highlighted throughout as a means of ensuring that the critical results of my research and overall opinions are clear and unmistakable. It is my imperative to present that which every citizen of the world needs to know most about our climate. It will be something you have not been allowed to hear for almost 20 years. It will be . . . the truth.

This book is about climatic and geological effects based upon changes in the Sun. It is not, however, about the ongoing debate over man-made global warming. It is my belief that the man-made, or anthropogenic, global warming (AGW) concept is a flawed theory based upon misleading, incomplete, and heavily politicized science whose time has come and gone, and will only be referred to in the context of pointing out its obvious myths.

In my own review of the United Nations' physical science reports and thousands of pages of climate science research and opinions from credible climate science experts around the world, I have concluded that becoming part of the ongoing global warming debate would be a waste of precious time. In short, my opinion on the topic is as follows:

> The theory of man-made global warming and climate change based on human greenhouse gas emissions is the greatest international scientific fraud ever perpetrated on the world's citizens!

On the question of what we should be using as a guide to determine how and when climates change, I offer this foundational and direct answer:

With regard to whether one should evaluate climate changes using the man-made climate change theory involving greenhouse gas emissions or the natural cycles of the Sun, the answer must be based on proven results.

Since the advent of industry, the concept that mankind's industrial greenhouse gas emissions causes climate change has never shown itself to be a reliable predictor of climate change. On the contrary, it has been routinely wrong. Further, recent developments have shown that the anthropogenic global warming theory may have been based upon faulty and manipulated data, driven in part by political motives rather than reliable scientific rationale. This has led to unsound and misleading conclusions and predictions by the theory's leading advocates.

On the other hand, the RC theory and similar theories by scientists based upon well-established and proven solar cycles — cycles that have been consistent for hundreds, if not thousands, of years — have shown themselves to be accurate to over 90 percent.

Given what we now know, there is only one reliable approach for climate change prediction: that which is based primarily on the Sun's activity, influenced by the other planets of the solar system and their combined effect on the Earth-Moon system.

We should not be surprised to see record-setting, catastrophic volcanic eruptions and earthquakes taking place in the next two decades, if not sooner. In fact, research shows with a high degree of certainty that they are coming.

I know what I am saying is a lot to digest — perhaps even shocking. Preferably, there would be time to develop this information and disseminate it gradually, over many years, so it could slowly filter into and become part of a body of understood and accepted thinking within the scientific community, the media, our educational systems and governmental organizations, and even in Hollywood movies and documentaries.

A key point in this book is that I will not try to convince you of my forecast by use of phony lines like that of the anthropogenic global warming movement that state, "Just wait until the year 2100; you will see then that we're right." I believe you will see here and now that, since the year 2011, our climate has already changed.

In appendix 2, Leadership in Climate Change Research, I have provided a chronological record of the US leadership role — initially by myself and then later under the auspices of the Space and Science Research Corporation (SSRC) — in publicizing the science behind this next climate change to a destructive, cold period. You might even want to start reading there, for it provides an overview and outline of the official record, via letters and press releases issued to our government and the media, of what is heading our way. In March of 2011, the former Space and Science Research Center was reorganized as the Space and Science Research Corporation. With this name and structure change, I was joined by some of the world's leading scientists to form a group of supporting researchers to help the new SSRC achieve its mission.

The facts and details in the press releases included in appendix 3 are voluminous. These releases were sent to many political leaders, media personalities, corporate heads, and AGW zealots.

One could easily write another book covering the stories behind that one appendix. It would include my attempts to expose the clear case of fraud within the US government on the subject of climate change policy development and one of the most reprehensible financial scams of the century: carbon credit trading.

Still, another book could be written on how propaganda, confusing words, and seemingly reasonable dialogue has aided AGW adherents in their plan to classify any kind of extreme weather or climate fluctuation as being caused by mankind's greenhouse gasses. To assist you in getting through this morass, I've included a glossary of my own definitions covering the past and present era of climate change.

I believe our discussion for the next few decades should be redirected away from the pointless debate over man-made climate change to the amazing and dramatic changes taking place in the Sun — the supreme natural power in our solar system. It is like the late Paul Harvey would say: "Now it's time to hear . . . the rest of the story."

The Sun that has warmed us to record levels in past years is now reversing course. Our lives will soon become unavoidably more difficult because of it. This book is written to give you and your loved ones a heads-up — a warning to mitigate, if not avoid, much of the coming difficulty. An incredible solar hibernation has begun in our lifetime. Our warm Sun is now becoming a cold Sun. I recommend that every American, every citizen of this planet, get ready for the cold.

The story you are about to read will not be just a dose of new scientific theory or misguided climate change activism. It will be perhaps the most important climate story — the most important human story — of the twenty-first century.

1

Moment of Revelation . . . WOW!

"And God said, let there be light; and there was light."

— Genesis 1:3

MY BREATH WAS TAKEN AWAY in an instant. I made a slow, deliberate backward rock in my chair from my worktable and uttered a nearly-silent, "Wow!" And again, "Wow!" After the longest, deepest inhale and exhale, I whispered the words, "Surely, this cannot be!"

I regained my train of thought and looked at the data another time, and twice more just to be certain. There was no question now. The data was solid; the conclusions like a rock. I thought, *Had I just unlocked what may turn out to be the answer to one of the most perplexing global climate questions of the modern era?*

What I had uncovered was both inspirational and gut tightening. Extrapolating from my charts of data, I had discovered that a coming climate change would lead to a period of potentially disastrous cold. Further, it looked like it was going to start its downward spiral within the next three to, at the outside, 14 years! My calculations said three years, but I realized we needed time to prepare. My hope was for 14.

After weeks of concerted, sunup-to-late-night effort, poring over hundreds of reports and thousands of pages of research, articles, tutorials, and reference material, I was convinced that I had just found something unique — something that everyone would want to know. For decades, we of this baby boomer generation have been perplexed — transitioning from global cooling concerns of the 1960s and 1970s to the global warming scare of the 1990s and the new millennium. Now, I've found that we're heading back to the cold again, but this time a very different kind of cold. What I stumbled upon was a particular cycle of the Sun's activity — one of its most important — that regulates when the Sun heats and then, by lessening its intensity of radiated energy, cools the Earth. Except this time, I found, the cooling is expected to be extreme. My research findings became even more ponderous as I slowly began to consider their potential impact. At once it became clear that this news might prove a resounding scuttling of the man-made global warming theory. Now, the opposite scenario for climate change, a major cold period — perhaps a dangerously cold one — was about to envelop the Earth. I was not to learn just how cold until after another three weeks of research, and when I did, I was shocked again.

> Simply put, there is no one alive who has experienced the depth and extent of the cold that will soon descend upon us!

But what about the prediction of when all this would start? As soon as three years? Who could believe such a forecast? Twenty years of global warming propaganda and UN predictions of ever-increasing temperatures year after year until 2100, and now here I am, out of the blue, essentially unknown to the scientific community and not one scientific paper to my name, making such a grand pronouncement. Regardless of my space program and high-tech background, I am neither a meteorologist nor a climatologist, but there I was, about to tell the world that Earth's climate is going to reverse course within three years! I said to myself, *No one is going to accept such a preposterous proposition! You have got to be kidding! You're not really going to come out with such a story, are you? Are you?*

The hour after my "wow" moment was humbling . . . totally. To get a grip on this discovery, I took a break and walked downstairs and out the front door of our home, a small townhouse on a beautiful cypress- and oak-lined golf course near Orlando, Florida. The warm, spring day had a slight breeze — a bit cooler than normal for the time of the year. And then I thought, *How appropriate.* Later I was to learn that, sure enough, April 2007 temperatures in the United States were slightly cooler than the twentieth century mean.[1] It would be a year later before I would see just how cool 2007 had gotten.

My pace was slow and deliberate, filled with the import of the event and its predictable aftermath. Questions began to flood my brain: How will this be received by the scientific community? By global warming advocates? By the government? And most importantly, what can I do to alert the people? And then there was the crucial commodity of timing. It seemed as though I was forever too far ahead with my goals and the technology at hand. For me, I guess I wasn't happy in any past job unless I was tackling the toughest technological challenge I could find. This time it was different. This was not the next-generation rocket, advanced language processor, space flight training center, or global communications system that was going to reform an industry (if not create a new one). This time everyone on Earth — every neighbor, every friend, every relative, every government official, every scientist, every person, rich and poor alike — was going to be affected.

Fifty paces into my post-epiphany walk, I turned the corner to gaze upon a wide open field, perhaps 400 yards by 200 yards and bordered in the distance by a great stand of bushes, tall oak trees, and huge cypress that lined the lake beyond. It is more than a hundred acres of still undeveloped land that I had tried (unsuccessfully) to get local and state government to purchase and turn into a nature preserve. Now, I could see the land being used to build apartments and condos to house the greater numbers that would want to move south to avoid the coming cold. Coincidentally, not three weeks earlier, members from our homeowners association attended the presentation of yet another developer who wanted to show us his firm's plans for a new condo project on the land. The baby boomers like me and my wife are already nearing retirement, and the mother of all retirement floods is already supposed to be coming to Florida over the next two decades.

A milestone event of the retirement issue occurred on October 15, 2007. On that day, the first official baby boomer, Kathleen Casey-Kirschling (no relation), filed for Social Security benefits. She is the first of 80 million retirees who will start drawing from the retirement fund (or what's left of it after the 2008 Wall Street debacle).[2] She may also be one of those who has already planned to move south to Florida. What will the rapidly advancing, prolonged cold period do to accelerate this migration? How will an already overdeveloped Florida handle the deluge of those fleeing what for them would probably be the most bitter and unrelenting cold of their lives — one that could last for 20 or 30 years?

I had spent many days birding on this still natural, undisturbed setting, which I was concerned would be paved over. With my Indiana Jones fedora, a pair of heavy binoculars around my neck, a notepad, and a copy of Roger Tory Peterson's bird book, it was my way of enjoying nature — literally counting my blessings and getting away from it all. When you have nesting bushes, tall trees, open pasture, and water all in the same area, you have one of the best possible environments for birding. The variety of species in such a setting can be impressive. Was that enjoyment also coming to an end? Surely Florida would be spared the brunt of the cold, wouldn't it?

Now another line of questions rushed into my head. What would be the impact on wildlife, specifically birds, their migrations, and their outright survival? Many species of migrating birds mate and raise new young in the northern United States and Canadian wilderness. Once reared and ready for travel, they make their way south for the winter. A favorite sight of mine is the sandhill crane. These beautiful, gray-feathered, long-legged, red-capped, graceful flyers are among the best known migratory birds in North America. They have already adapted to the ever-shrinking Florida environment and are routinely seen near highways and even in the backyards of homes built on the many thousands of lakes and ponds we have in Florida. What will be their fate when the long years of seemingly unending cold are upon us, or, in a worst-case scenario, when there is no spring or summer, like what happened in 1816, the so-called "year without a summer"?[3] Will it be that bad? I had to know. I would have to go back to the data. I needed to know just how cold it was going to get. I turned back toward the house. There was more research to be done.

The initial insight into what was to become a new theory for the Sun's heating and cooling of the Earth was like pulling on the single thread that leads to a larger unraveling. In the course of reviewing data obtained from Internet searches for another book altogether, I was particularly intrigued by a couple of charts of sunspot activity. They immediately struck me at once as having some underlying periodicity.

Sunspots had been recorded for centuries, but Galileo was one of the first to correctly interpret them shortly after he improved the telescope.[4, 5] They have since become recognized as harbingers of the Sun's activity level and have been studied extensively for 400 years. From this study, various cycles of the Sun have been determined. The best known is the 11-year solar cycle, also called the Schwabe cycle. This 11-year span is an average duration; it actually varies, with some cycles as short as 7 years and some as long as 17. During the Schwabe cycle, the number of sunspots reaches a maximum, and then drops to a minimum, and then reaches its next high point 11 years later (again, on average). There are many

SUNSPOTS – APRIL 25, 2011

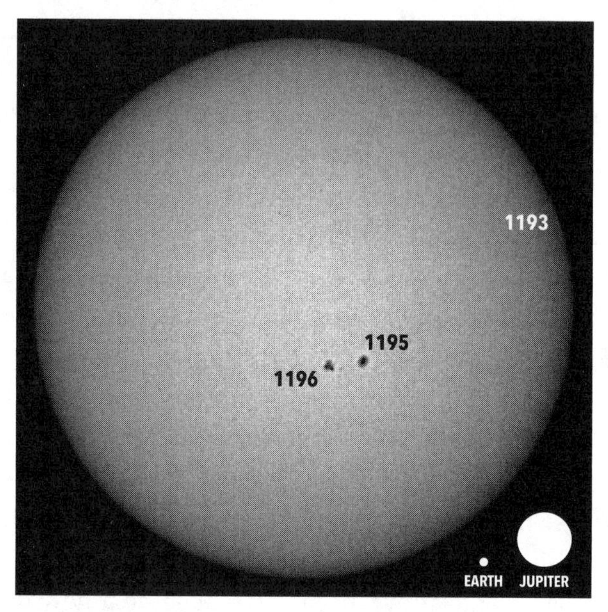

Source: NASA/ESA SOHO Satellite

Figure 1-1. Sunspots. This April 25, 2011, photo of the Sun, which remains in its solar hibernation mode, shows a few lonely sunspots. A comparison of their size with the Earth and Jupiter are shown in the lower right.

other solar cycles, and most are much longer, far more powerful, and driven by the intricate movements of the Sun, Earth, and the other bodies in our solar system.

Most of us take them for granted, yet these natural cycles and periodicities of the Sun, or "oscillations," as researchers call them, bring us light and darkness, warmth and cold. Some cycles are on the order of a few hundred years; others stretch from several hundred years to thousands, tens of thousands, or hundreds of thousands of years.

One of the longest cycles is that of the ice ages. In between a cycle of every 100,000 or so years of essentially an icebound Earth, we have what are called "interglacial warm periods." For the past 11,000 years, we have been living in one of these rare interglacial periods, called the Holocene warm period.[6] We humans have found our planet hospitable enough during this era such that we could move quickly out of the caves that sheltered us from the last ice age and into "recorded" history for the first time. We have thrived and multiplied. Boy, how we have multiplied!

We all have our own apportionment of gifts and flaws. Allow me the author's prerogative of steering clear of my many flaws and past miscues for now. Instead, let me tell you about one fortunate gift of mine that came into play that spring day. It is in pattern recognition. I could immediately see a pattern in the long-term sunspot charts. The first eye-catching relationships were from charts showing sunspot records covering the past 400 years. From that point on, all else followed a logical path of deduction and reasoning that weeks later produced what I would come to call the Theory of Relational Cycles of Solar Activity, or simply the Relational Cycle (RC) theory. In the course of developing the theory, I was to uncover astounding, yet previously obscure and little-known findings on the cycles of the Sun that would be news for the general public and many scholars as well.

In the specific field of science that I was studying, solar physics, I was also to learn, with subsequent research to confirm my findings, that the theory was always there waiting for someone to come along and put it all together. Many outstanding researchers, in fact, had gotten close to doing just that. Some had already discovered the same individual solar cycles and given them names in honor of leading scientists, including Wolfgang Gleissberg, de Vries, and Hans Suess. Others, like me, had further predicted a coming cold period. But no one had put it all together, as I had, into a working theory, and no one that I could find, at least in the United States, had taken the science to the next level in predicting the next series of climate changes. Most importantly, no one else was on a mission to get the word out about the difficult times ahead.

So what was this discovery? What revelation was it that was going to plunge the world into a long cold? The discovery was, in essence, that among a group of relatively short cycles of the Sun, there was one specific cycle — a 206-year cycle that I came to call the Bicentennial Cycle — that was the primary scheduler for climate changes on Earth on a scale of many decades.

The Bicentennial Cycle of 206 years correlated with near 100 percent accuracy to every major cold-temperature period over the past 1,200 years. Based upon my calculations, I discovered the next cycle change was imminent, and because of the record cold it would bring, I quickly realized this could mean a rough, perhaps dangerous period for many of Earth's inhabitants. This linking of the solar cycles to Earth's temperature, coupled with the prediction of the next climate change being a potentially dangerous cold era, was the crucial finding. This was the "wow!" that struck me at 2:00 p.m., April 26, 2007. It all came together in an instant.

This was not the typical scientific finding that would require more research and perhaps decades of waiting to see if it would take place. The Bicentennial Cycle was one that we mere mortals could directly experience during the lives of our parents, our own lives, and those of our children (simultaneously, of course). These are the types of cycles of the Sun described in the RC theory — those that have real meaning and tangible impact on our lives. While, from a scientific standpoint, the research into cycles that are a thousand or more years long can be quite interesting, unless one of these major cycles is also about to turn over, they have no relevance to us in our lifetime, much less our day-to-day existence. The 206-year cycle and other smaller cycles I found are the ones that have real impact on us. These are the ones that we can relate to, as they fully account for the puzzling swings of global climate change that have perplexed us all for the past two centuries. If one can accept that major solar activity minimums have been in lockstep with significant cold periods as the physical records show, and that nothing man can do will change the cycles of the Sun, then one can only conclude that the next solar hibernation will also bring with it a calamitous cold era as it has done before.

This is the central premise behind the forecast of difficult times ahead. This straightforward logic appears inescapable regardless of one's prior beliefs about AGW or any other climate change theory. If you can agree with this one paragraph above, then you will likely find yourself abandoning the now-discredited concept of AGW and its easily disproved forecasts of ever-increasing global temperatures.

Within a few weeks after this most important Bicentennial Cycle discovery, I formulated the Theory of Relational Cycles of Solar Activity to account for its effects and those of other similar solar cycles. Here, then, is that theory and its seven main elements that came from my independent research. It is my fondest hope that with the growing support the theory is receiving from top researchers and scientists from around the world, the next global climate change, which will be a return to a deep and prolonged cold period, will nonetheless be met by a people well-prepared to endure it.

The Theory of Relational Cycles of Solar Activity (the RC Theory)

- There exists a family of solar activity cycles that has a profound and direct influence on Earth's climate.
- These cycles are called "Relational Cycles," since their effects can be experienced, or related to, during one or two human lifetimes.
- There is a "Centennial Cycle" of 90 to 100 years' duration, which manifests itself in minimum solar activity and associated low temperatures, with episodes lasting a few years to one to two decades.
- There is a "Bicentennial Cycle" of about 200 years that is the most powerful of the Relational Cycles and has significant effects on the climate of the Earth, lasting several decades

and resulting in the most extreme variations in solar activity and in Earth's temperatures.

- These cycles are correlated strongly to all past major temperature lows.
- There is remarkable regularity, and hence predictability, among these oscillations, such that the theory may be a powerful tool in forecasting major temperature and climatic cycles on Earth, many decades in advance. There may be other Relational Cycles of shorter duration accounting for lesser solar and climatic events, which may be revealed in subsequent research.

Note the 206-year length in the Bicentennial Cycle. My first calculations showed it at 207 years, and such was the number I first publicly announced in my initial press releases of April and May 2007. Other researchers had also found the cycle length at 207 years. A more refined calculation for my subsequent scientific papers resulted in a 206.25-year cycle period. It was an improvement, albeit of less than 0.4 percent. For accuracy thereafter, I began using the 206-year figure. The error probability around the 206-year number, however, is sufficiently large, and other researchers observed it at 200, 202, 206, 208, and 210 years. Any number in the area of 200 years is now widely accepted as the same solar activity cycle. What I had independently found and named the Bicentennial Cycle was in fact discovered many years earlier and named the de Vries or the Suess cycle, though I was completely unaware of this prior to the completion of my own independent research. Similarly, the 90- to 100-year Centennial Cycle is also called the Gleissberg cycle.[7]

During the course of my corroborative study, I read many other research papers and abstracts that came to the same conclusions by differing techniques, each of us providing validation for one another's work. This is a common scientific approach when trying to understand our world: coming at a problem or question from various angles, and yet arriving at the same conclusions. In this manner, the many reasons for discounting or supporting a particular theory rest upon the most refined, comprehensive, and objective scrutiny.

The sixth element of the theory is hugely provocative — what technologists call "disruptive." Controversial new theories, advanced new technology, and milestone innovations come along rarely, but when they do, their impact on their field (science, engineering, industry, etc.) can be so overwhelming that they disrupt the existing scientific, economic, or technological order. They dramatically and quickly challenge the status quo and break down barriers of conventional thinking. I could see that as soon as I wrote this theory, particularly the sixth element, it clearly fell into the disruptive category. It was abundantly clear to me that the nature of its precise importance was absolutely without question.

Curiously, even as I write this book, I am awestruck by newspaper articles which, quoting several scientists and experts on climate change, say there is still no way to predict major climatic events . . . even a year ahead. Now with one sentence of 35 words, I knew all that was about to change — not just for that day, or the next few years, but for decades, if not hundreds of years or more. And they are going to stay that way, until open scientific research, combined with a renewal of freedom of speech in the scientific community, produces a better theory (or a better 35 words).

I knew right away what needed to be done once the theory, and especially its ability to roughly predict the future, was firm in its form. We were in for perilously bad weather for the next few decades, especially because, until my research and that of others was unified and I was able to formulate (and name) a solid theory, the world has been unprepared to deal with such a situation. The initial steps were certain, and I took them into hand immediately. The word had to get out.

Therefore, on April 30, 2007, I published my first press release to warn the United States that the next climate change was coming. (See appendix 3.)

2

What Happened the Last Time?

"Those who cannot remember the past are condemned to repeat it."
— George Santayan

THE OFTEN-QUOTED STATEMENT BY George Santayana bears repeating. Whether it is the learning process of each new generation that comes along, or that prior generations simply forget, we continue to find this sage advice befitting over and again. A review of past Bicentennial Cycles shows that solar history, like human history, repeats itself. In this chapter, we will look into what happened the last time a solar hibernation took place, in hopes it will give us some idea of what we can expect during the one that has just started. As you will find out in this chapter, the Sun's cycles are strongly intertwined with our own. You will find in this chapter just how closely the human species is tied to the Earth-Sun relationship. For example, here you will learn for the first time that the United States of America probably owes its very existence as a sovereign nation in large part to a cycle of the Sun!

The last time a major solar minimum — a solar hibernation like the one about to hit the Earth — happened was between 1793 and 1830; this was called the Dalton Minimum (DM). The period was named for the famous British scholar and researcher John Dalton. Dalton (1766–1844) was a chemist and mathematician who was also interested in meteorology. He often made notes and kept records about the weather, especially during the time that would later carry his name. He is best known, however, for formulating the structure of the atom and the atomic theory.[1] Figure 2-1 shows the sunspot cycle from 1610 to 2000. The Dalton Minimum is highlighted.

Some scholars use 1795–1825 as the DM period. I have chosen to begin the DM at the solar cycle "4 Prime," discovered by I. G. Usoskin, K. Mursula, and G. A. Kovaltsov.[2] In their paper "Lost sunspot cycle in the beginning of Dalton Minimum: New evidence and consequences," they make a convincing case that the latter part of cycle 4 was, in fact, a small solar cycle with very low amplitude, hence 4 Prime. I end the DM in the year 1830, which

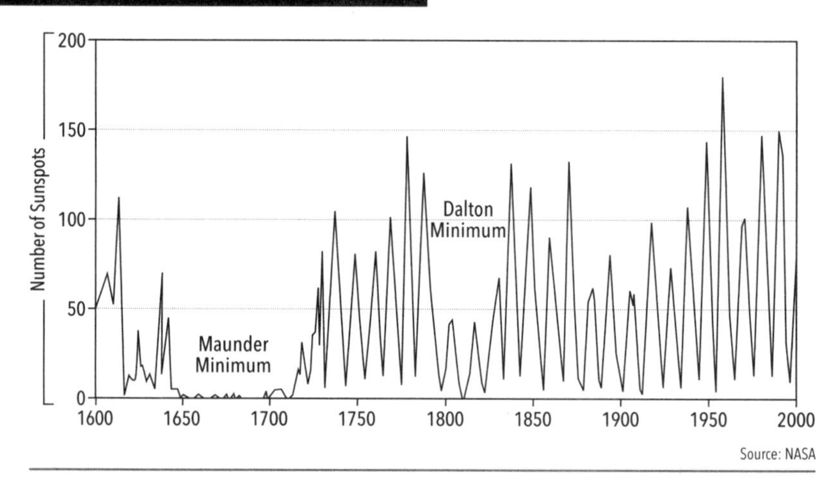

Figure 2-1: Sunspots from 1610 to 2000 showing Maunder and Dalton Minimums. Each 11-year, solar cycle is given a number, starting with number 1 in the mid-1700s. We are now in cycle 24, which started in 2008. The Dalton Minimum began in cycle 4 and ended after cycle 6.

GRAPH OF SOLAR CYCLE 4 AND 4 PRIME

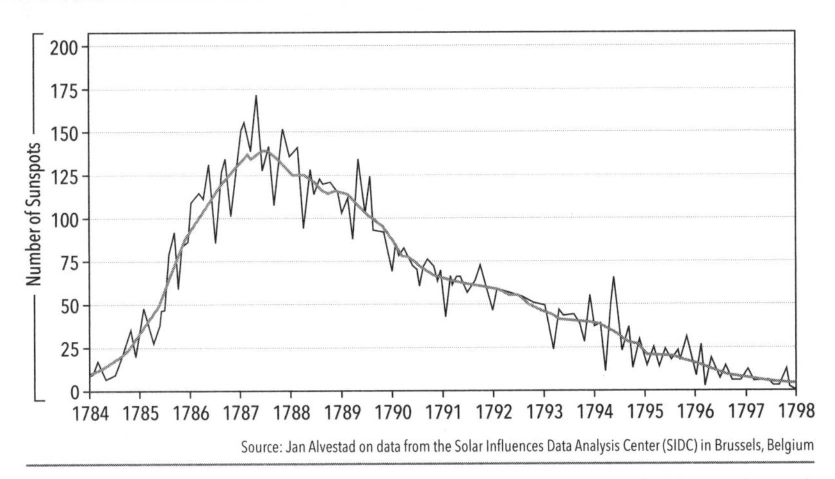

Figure 2-2: Note the period from 1793 to 1798. This is the period identified by Usoskin et al. as 4 Prime, the "Lost Cycle." I concur with their findings and have used 4 Prime as my start to the Dalton Minimum.

was where the solar cycle 7 began to exceed the three previous low-sunspot cycles, namely 4 Prime, 5, and 6. Figure 2-2 shows the cycle 4, with 4 Prime extending from 1793 to 1798.

My prediction for the current solar minimum and associated cold period is similar to (or more severe than) what was experienced during the DM. There are crucial differences, however, between the DM of the past and the next solar minimum. These differences fundamentally change the scope and depth of the ill effects that will strike during the coldest years of the coming solar hibernation. They include the following:

- We live in a world today that is vastly more interdependent, with nations on one side of the planet heavily reliant upon other nations on the opposite side, particularly for much of their food and energy stocks.
- We are intimately tied to new technologies and the power systems that keep them running, both of which are likely

sources of failure during weather extremes, and in the process can affect millions of lives. During the last DM, there was relatively little dependence on technology or power compared to today.

- Most people on Earth are now urban dwellers, with a small percentage of farmers that feed the world, far fewer than we had in 1800. In other words, most people on Earth no longer grow or raise their own food, as was the opposite case in 1800.

- We have 7 billion mouths to feed in the world today[3] compared to the 1 billion we had in 1800.[4] We will be up to 8.25 billion — another 1.25 billion at the dinner table — by 2030, a short 20 years away.[5]

- The DM was preceded by a solar maximum that contributed to substantial global warming at that time. The same is true today, although the Bicentennial Cycle of modern times, the "modern maximum," is far more intense.

- The decline from the past Bicentennial Cycle peak into the DM resulted in major agricultural and temperature disturbances. Yet, the potential for even more disruption exists, because my prediction (and that of other scientists) is for a drop from the highs of the current Bicentennial peak into the next minimum — again, a solar hibernation — that could be much more steep than the DM. Highly variable and strong weather patterns that may set both warm and cold records in a short span of time may take place during the transition period to a predominantly colder world.

The current state of the world's understanding of climate change will work directly against early warning and preparation for the next solar hibernation. This is a direct result of the one-sided focus on AGW, which, unless a correction of thinking is made, will delay, if not prohibit, an effective response to the new climate era. Early warning, much less action to reduce the effects of cold weather of this kind, was unthinkable in 1800.

What was life like in the United States between 1793 and 1830? It was quite different from today, without a doubt. There were only 5.3 million people in the United States. (We have more than three times that number in Florida alone today.) There were only 16 states, and most of what was west of the Mississippi was undeveloped, except for the West Coast, specifically along the coastline.[6] People traveled on horseback or by buggies or wagons. There was no such thing as electricity, lightbulbs, cars, or trucks. There was no television, no radio, no cell phones, no paved roads, no airplanes or airports, no trains, no busses, no modern hospitals, and no computers.

In fact, that era could be better described by what people did not have rather than what they did when compared to our fortunate, high-tech times. The vast majority of people lived off the land. Most homes were heated from basic fireplaces and wood-burning stoves. The US Constitution had only recently been ratified in 1787. John Adams had just been elected the second president of the United States, after George Washington, and Thomas Jefferson would soon follow as the third president.[7] Europe was in turmoil (again), with Napoleon about to make the fatal blunder of invading Russia and then trying to survive one of the worst winters in the history of warfare. No one was around to warn him of the worsening cold era that would later be called the Dalton Minimum. When he invaded Russia, he had 600,000 soldiers. When he finally retreated, at least 400,000 had perished — many from the terrible cold of the winter of 1812–1813.[8] Most had died in battle or left the field, but many simply froze to death. During the DM, tensions were growing between the United States and England, and war between them was ready to erupt again.

Here is but a brief synopsis of mostly well-known events that took place just before and during the DM as viewed primarily from the US historical perspective:

- 1787: The US Constitution ratified by the states.
- 1789: George Washington elected as first president of the United States.
- 1789: The French Revolution begins.
- 1793: Dalton Minimum begins: Solar Cycle 4 Prime.
- 1797: John Adams elected second president of the United States.
- 1801: Thomas Jefferson elected third president of the United States.
- 1803: Lewis and Clark begin to explore the northwestern United States.
- 1803: The Louisiana Purchase negotiated with France.
- 1807: Robert Fulton launches his steam-powered boat, the Clermont.
- 1809: James Madison elected fourth president of the United States.
- 1811–1812: The New Madrid earthquake strikes the Mississippi valley; the first quake occurs on December 16, 1811. It is the most powerful series of earthquakes in North American history — a series of three 8.0 temblors, plus many smaller ones.
- 1812: The War of 1812 between the United States and England begins.
- 1812: Napoleon invades Russia and suffers massive losses because of bitter winter weather.
- 1815: The Mount Tambora volcano erupts, April 5, 10–11. It is the largest and deadliest volcanic eruption in recorded history at the time, claiming 90,000 lives.

- 1815: Napoleon suffers loss at Waterloo on June 18.
- 1816: The "Year without a Summer." Bitter cold weather hits New England, and the destructive frost spreads as far south as Pennsylvania.
- 1816: In May, a frost hits from New England down to Virginia, and in June, people go sleighing after a freak snowfall.
- 1816: On July 4, Independence Day, another killing freeze strikes, with more snow and ice reported in Virginia.
- 1816: In August, frosts and snow strike New Hampshire, killing off what few crops still survive. Two months earlier, temperatures had been in the 90s.
- 1816–1823: Hundreds of thousands die, possibly as a result of cholera that spreads from India to New York City, related to regional conditions from Mount Tambora's eruption.
- 1817: James Monroe elected fifth president of the United States.
- 1825: John Q. Adams elected sixth president of the United States.
- 1830s: A second wave of cholera strikes Europe, also possibly related to the Mount Tambora eruption. Hundreds of thousands more die, especially in France.
- Thousands die in New England from the cold and aftereffects, and thousands leave for Indiana and Illinois. The migration may have been a key factor in these areas becoming new states of the newly formed United States of America.
- The combined cold and heavy rain damage of 1816 causes potato, corn, and wheat crops to fail across Ireland and England and, along with collateral typhus outbreaks, thousands more die.

- Rapid temperature fluctuations are common and extreme, with reports of 90°F and higher temperatures plummeting to near freezing in a matter of hours!
- Crop prices skyrocket. For example, prices for oats, essential for horses (the main mode of transportation), go up 700 percent.

The world was a different place indeed. Just the process of communication took long periods of time. News was carried between nations via sailing ships, extending global communications to months before word of any significant event could be passed on. We all remember from high school that in the War of 1812, for example, Andrew Jackson fought the British in the Battle of New Orleans without knowing a peace treaty had already been signed, ending the war.[9]

Volcanoes and Earthquakes

The world's largest volcanic eruption in recorded history took place in April 1815 on the island of Sumbawa, Indonesia. The eruption of Mount Tambora was to become a global climate-changing event that in the United States would lead to 1816 becoming the "Year without a Summer." In America and Europe, however, news of the cataclysmic event was delayed for months, since the telegraph and telephone had not been invented.

Robert Evans does an excellent job summarizing the Tambora eruption and its effects on global weather, especially in the United States, in his piece for *Smithsonian;* he also describes his trip to present-day Tambora.[10] Combined with information compiled in Wikipedia[11] and the work of Dr. Willie Soon and Steven Yaskell, the Year Without a Summer[12] has left in its now-subtle wake a number of reasons we should be very concerned about the solar hibernation that has just started. Let's look at some vital pieces of history from these sources about the last time this solar event occurred.

The eruption of Mount Tambora was to become the hallmark geological event for the global cooling of the DM that had started 22 years earlier. Its volcanic ash spread across the globe over the next two years. Murphy's Law was in full force. It wasn't bad enough that the solar hibernation was in full swing, but to make matters worse, a major volcanic eruption had to take place, blocking out even more of the precious warmth of the Sun. This eruption was ten times more powerful than the better known and documented Krakatau eruption of 1883. It was a hundred times more powerful than the 1980 eruption of Mount St. Helens.

Historically, the Indonesian islands have seen some of the largest volcanic eruptions ever recorded. Figure 2-3 shows the many past and currently active volcanoes of this archipelago. It would not be unusual at all for Indonesian volcanoes to become active during the next solar minimum and once again add to the deepening of the cold era.

Another superior reference is *Volcanoes in Human History* by Jelle Zeilinga de Boer and Donald T. Sanders.[13] This book

MAJOR VOLCANOES OF INDONESIA

Source: USGS

Figure 2-3: Major volcanoes of Indonesia with eruptions since 1900 A.D. Note the location of Tambora and Krakatau, sites of two of history's largest recorded eruptions in 1815 and 1883, respectively.

provides one of the best and most detailed accounts of the Tambora eruption and many others. It chronicles the weather-related effects of the destructive Tambora event and the spread of disease from 1817 to 1832.

The Dalton Minimum was an active period for volcanic activity; in addition to Mount Tambora, La Soufriere on Saint Vincent in the Caribbean erupted in 1812, as did Mayon in the Philippines in 1814.[14, 15] The question then arises: What will happen during the next solar hibernation? Will we also experience volcanic activity that will add to the solar cooling? The answer, most likely, is yes! At any one point in time, there are about 50 active volcanoes around the world, according to the US Geological Survey (USGS). We should expect to deal with multiple geological disasters, including volcanoes and earthquakes, during the next solar hibernation. I will provide more on this important, additional solar hibernation correlation later.

Even with the extensive research that has been done on the effects of Tambora on the DM, I still see facts on the table that say the Sun's lower activity in the DM was already taking global temperatures lower by 1816 and was the predominant force for lower temperatures during the entire DM period. These facts include the following:

- The DM, signified by dropping sunspot counts, began in 1793, not 1815–1816.
- The DM had already caused temperatures to decline 22 years before.
- The effects of Tambora may have lasted two to three years; the DM lasted 37.
- Other reports of weather conditions show that Tambora's effects were certainly multiregional but not necessarily global. This effect is not unusual. Among volcanologists, the location and type of eruption, variance in weather, and wind and

ocean currents are known to affect the global wind dispersal of volcanic particulates and gasses over a wide range. In a paper published in 1990,[16] J. Neumann discusses the impact of both the Tambora and Krakatau volcanoes on the Baltic region and shows, for example, that grain harvests and death rates in the Scandinavian nations seemed to be unaffected by Mount Tambora.

- European temperature measurements during the DM showed the decline in place well before the Tambora eruption, and in fact, the lowest temperatures came before 1816, based on the European temperature profile done by David Archibald in his work, *The Past and Future Climate*.[17]

Archibald's data also shows a pronounced temperature drop in Europe in the mid-1780s. This was most likely the result of wide coverage by ash plumes and gasses from the eruption of Laki in Iceland in 1783, one of the largest eruptions of its type in history. This eruption was actually not like the explosive, single-crater mountain type (stratovolcano) that we typically associate with volcanoes, but rather was along a giant fissure that opened up and eventually emitted gas and lava from a string of 130 craters. Before it was over, much of Iceland's livestock had perished, along with 25 percent of its population.[18] A case could be made that the DM should be redefined for Europe with a start date of June 1783, coinciding with the beginning of the eight-month-long Laki eruption. This would be an artificial extension to the actual solar decline, which as I stated earlier began in 1793.

So what about our times? Do we see similar volcanic threats appearing? Yes! Lest we forget, the Icelandic volcano Eyjafjallajökull erupted in early March and April 2010. This now silent volcano, similar in type to the Laki event, grounded flights over much of Europe because of the wide-ranging dust plume it generated.[19] Should Eyjafjallajökull trigger the nearby Mount Katla

MAJOR VOLCANOES OF ICELAND

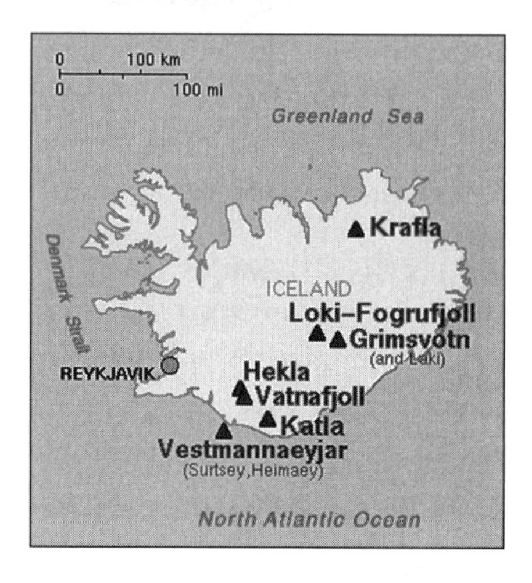

Source: USGS

Figure 2-4: Map of Iceland showing major volcanoes. On this map, Eyjafjallajökull, which erupted in April 2010 and shut down air travel across Europe, is just to the left of Mount Katla.

volcano, as it has a habit of doing, then another major European or even global temperature drop may be just around the corner.[20]

Without a doubt, major volcanic eruptions can have serious consequences for Earth's climate. When they occur during a solar hibernation, they can amplify the temperature drops that come with these long-term cold climate periods. *Even as this book nears completion, Mount Merapi in Indonesia threatens to blow its top. Over 200 people died during the first weeks of activity of this well-known volcano alone.*[21]

On the subject of supervolcanoes like Yellowstone, we should not be too concerned, since there are usually tens to hundreds of thousands of years between eruptions. However, if we suddenly see a spike in the number and intensity of earthquakes in the vicinity of the world's supervolcanoes, then everyone should start paying attention. As most are aware, these volcanoes can generate global geological catastrophes, resulting in incredible destruction, such as worldwide "volcanic winters" lasting for decades. If we see a rash of moderate to large earthquakes anywhere near these sleeping geological monsters, all bets are off. If one does erupt at full power, the world could be horribly transformed, with a large percentage of the world's population killed off within a year or two.

In addition to volcanoes, a brief review of available literature on the subject suggests that there is also a periodicity for much larger and destructive earthquake activity during solar hibernations. As previously noted, a series of massive, 8.0 earthquakes struck the general area around St. Louis, specifically New Madrid, Missouri, between 1811 and 1812, and were the largest such earthquakes in US history.[22]

These concerns were reinforced with the release of the Space and Science Research Center's preliminary report in January 2010, "Correlation of Solar Activity Minimums and Large Magnitude Geophysical Events." This report, which was completed before the devastating Haiti earthquake of the same month, is available at the SSRC website, SpaceAndScience.net. After several administrative delays unrelated to the research, the preliminary report was posted May 10, 2010 — four months before Christchurch, New Zealand, received major earthquake damage. It is a mind-grabbing set of findings that point to the potential for even greater global distress during the next climate change. Research by the SSRC shows that historically large volcanic eruptions and earthquakes are more likely to take place during solar hibernations. I believe major earthquakes and volcanic eruptions should be expected and planned for by all nations, especially those in geologically active zones that have a cyclical history with harmonic 100-, 200-, 300-, and 400-year intervals.

Given the high correlation of major geophysical events to solar hibernations based upon this study, I have notified the US Geological Survey; the major newspapers and governors of California, Oregon, and Washington State; as well as newspapers in the St. Louis area.

On Friday, March 11, 2011, the great Tohoku earthquake struck off the northeast coast of Japan, near the city of Sendai.[23] I watched Fox News and CNN accounts of the tragedy. Those of us watching the videos of the great wall of ocean water plowing inland from the resultant tsunami were aghast at the extent of damage that was unfolding before our eyes. According to the USGS, the quake's magnitude was 8.9 on the Richter scale, making it the fifth largest of the past hundred years. (It was later reclassified as 9.0.) The natural disaster claimed 20,000 lives and cost the country from $200 to $300 billion in infrastructure and property damage. Entire towns and communities along the northern coast of Japan's main island of Honshu were quite literally wiped off the map. According to Fox News, some coastal cities experienced such pronounced ground subsidence that they are now facing the bleak prospect of having to permanently abandon some areas where the water will not return to the sea. There was even considerable damage in Tokyo, and according to all major news sources, millions were without power, all airports were shut down, and trains and other commuter services around the country were interrupted.

The situation at Fukushima Daichi nuclear power plant on the coast of Japan near Sendai, as covered by numerous sources worldwide, saw four of its six reactors disabled, with three in a meltdown stage of some kind. The 12-mile exclusion zone around the plant was reviewed for expansion damage, with nearly 60,000 forced to abandon their homes and find shelter in nearby prefectures. Several reports estimated anywhere from many months to several years to bring the radiation danger under control.

During that time, I e-mailed Dr. Fumio Tsunado, one of the SSRC supporting researchers in Japan, to determine his status and safety. I was much relieved when he e-mailed back an hour later, saying he and his family were okay. He expressed his new concern that the quake could have reactivated a fault line in the vicinity of Mount Fuji, one of Japan's most dangerous active volcanoes. I asked him to keep me posted, especially since Mount Fuji is one of the volcanoes on the SSRC watch list; it has a bad habit of major eruptions during solar hibernations.

The impact of the Sendai, Japan, quake is far reaching, and once again it shows that what the Sun does for Earth is much more than just affect its climate. Following up the May 2010 prediction of major earthquakes and volcanic eruptions, I issued another timely release that stressed that this tragic quake was but a sign of more to come. (See appendix 3.)

We tend to view events in our life as isolated, but they are rarely so. I believe most important events are often the culmination of multiple, seemingly unrelated factors that combine to create a critical happening in history and our everyday lives. That certainly applies to solar hibernations, with their multiple effects of extreme cold, earthquakes, and volcanic eruptions. Add to that the potential for widespread loss of human life, and one has all the ingredients for historic events to transpire.

A Solar Hibernation Helped Create the United States

I grew up in a relatively conservative era. I went to traditional elementary schools, high schools, and college, and I heard about the Founding Fathers, the War of Independence, and the War of 1812 in terms of what took place and when, at least according to the history books. I learned many of the basic facts and about key individuals who influenced the course of our country's progress. In 2007, however, a new factor came into view following my research into solar activity cycles; this factor served to further revise my view of history and how closely we are tied to the natural world. Most of you will read about it here for the first time.

You are probably aware that if it had not been for French support during the War of 1812, the United States would have lost to the British, and our country would not be what it is today. World history would have been forever changed, to who knows what outcome. But there is more to the story. The French government that came to our aid had been born a mere 23 years before, following the French Revolution of 1789. You may remember from history class that the French Revolution was prompted in part by starving peasants who descended upon Paris and stormed the infamous prison, the Bastille. The peasants were starving because, for several years in a row, their crops had failed from excessive heat and lack of rain. There was little or no wheat, therefore no bread, hence unstable social conditions and the revolt. But during this same period of heat, there was also record cold! Here is a brief report of conditions of the time in a summary from *Encyclopedia Britannica*:[25]

> During the momentous political events of 1788–89, much of the country lay in the grip of a classic subsistence crisis. Bad weather had reduced the grain crops that year by almost one-quarter the normal yield. An unusually cold winter compounded the problem, as frozen rivers halted the transport and milling of flour in many localities. Amid fears of hoarding and profiteering, grain and flour reserves dwindled. In Paris the price of the four-pound loaf of bread — the standard item of consumption accounting for most of the population's calories and nutrition — rose from its usual 8 sous to 14 sous by January 1789. This intolerable trend set off traditional forms of popular protest. If royal officials did not assure basic food supplies at affordable prices, then people would act directly to seize food. During the winter and spring of 1789, urban consumers and peasants rioted at bakeries and markets and attacked millers and grain convoys.

But why had the crops failed? The crops failed because, at the time, there was in place global warming and drought, which I believe was caused by the Bicentennial Cycle, which was then at its peak of warming! That's right: the global warming of today is simply a repeat, if not a more severe one, of the same solar cycle and the same peak heating event that occurred during the late eighteenth century that helped spark the French Revolution. Similarly, we had record warm temperatures in 2010 and yet a dramatically colder, record winter in 2010–2011, both of which caused tremendous crop failures.

It appears from my research that the United States of America owes its very existence, in great part, to the natural reaction of the French people during the late 1780s to the side effects of drought and crop loss brought on by the last global warming peak and following extreme cold caused by the global climate changeover of the 206-year Bicentennial Cycle of the Sun! The loss of wheat crops and resultant shortage of bread were part of the reasons for the French Revolution of 1789. The government that eventually came into power would then come to the rescue of the United States in its war against the British in 1812. This vital support tipped the scale in favor of the United States, who won the war and went on to be the great nation it is today. If the Bicentennial Cycle of the Sun had not destroyed the French crops during the late 1780s, there might not have been a French Revolution, no Napoleon, and no United States of America.

This remarkable story of how nature and human history are intertwined does not end in 1812. The history of France and the world was to be affected once again by the Bicentennial Cycle. When Napoleon lost so much of his army in Russia during the terrible winter campaign of 1812–1813, it was within the flip side — the cold side — of the 206-year cycle. His defeat in Russia, caused in great part by a brutal winter, was in the middle of the DM, which was near its bottom coldest period. Would world history have unfolded differently if Napoleon had a science advisor with knowledge of the relational solar cycles, who would have warned him how bad the winter of 1812–1813 was going to be and advised that he conduct his war campaign with their effects in mind? It is a curious speculation, but one which is relevant today. Will world leaders follow the wisdom of Santayana and the lessons of Sun-related climate change during the critical period of

NAPOLEON'S WITHDRAWAL FROM MOSCOW

Figure 2-5. Napoleon's Withdrawal from Moscow. Painter: Adolf Northern

the history of France from 1789 to 1813, or will they once again ignore history and dare to challenge nature in a battle between politics and ego on one hand and the omnipotent Sun on the other? Any bets on who will be the victor in such a confrontation?

The Dalton Minimum cold weather bottom came in 1815–1816, some 26 years after the peak heating of the last Bicentennial Cycle around 1789, the year of the French Revolution. Note, if the peak of heating is found out to be in 2005, and we see a similar 26 years pass before the bottom of our own next cold period, that would make the next coldest period around the year 2031. That year just happens to be the same year I have calculated from carbon 14 data as being the next solar hibernation low point!

If the peak of the Sun-produced global warming turns out to be 1998 or the current warm year of 2010, as current temperature records show, then the bottom of the next solar hibernation could be as early as 2024 or as late as 2036. Again, my math says 2031.

Either way, it is coming, and the Sun will decide the exact schedule — not mankind!

One of the most frequent questions I get is: If there will be a change due to a coming cold period, why is it so darned hot? The near record-high temperatures of 2010 certainly demand an answer to the question. The answer can be found in the thermodynamic equations of the Earth-Sun system. With a few supercomputers and a lot of money and time, one might be able to resolve this relationship with greater precision and accuracy. In essence, there is a lag time between peak solar activity as measured by sunspot counts and Earth-related heating as measured by temperature. While much research is needed in this area to make firm conclusions, Earth may not be that different from a plate of food heated in a kitchen microwave. Most instructions warn that one must be careful after heating food and removing the dish because the food is still heating up, even after it's removed from the microwave. We see this same effect every day as the Sun reaches peak around noon, yet the warmest time of the day is delayed until two or three o'clock. Likewise, Earth may still be heating up even after the Sun begins its decline after a 200-year peak in solar activity.

If we look at the last time a solar hibernation took place, in the DM, we can see that significant heating occurred (witness the cause of the French Revolution) just before the dramatic DM temperature decline began. In our own times, we have seen record heat in 2010, just prior to my predicted rapid descent into the next cold climate era. The AGW community once again pointed to this year's heat as another sign of human-caused global warming. Not so.

The heat of the year 2010 is not a sign of global warming. It is, however, a sign that we have had 13 years since the record year of 1998 without any effective growth in the world's temperature profile. Global warming has certainly stopped and, according to long-term trend lines, is already heading down, having peaked, by my calculations, between 2005 and 2007. During the last two decades, the world's citizens have been subjected to a flood of media reports about global warming, along with the current US government and UN officials pushing the corrupted AGW theory. There has been an intensive international campaign aimed at convincing us of a threat of ever-increasing global temperatures from human-caused global warming. It is, in retrospect, a great historical contradiction that over this same period of time there has been no global warming.

The past is clear about the effect of solar cycles upon the Earth's climate. For those who do not wish to be "condemned to repeat" the past, we must first remember a critical lesson from the last time a solar hibernation struck. Had he known of this during his time, Napoleon Bonaparte's reign and the history of the world could have been quite different. And what is that lesson?

It's always hottest before the cold.

3

What Do We Do Now?

"What do you expect us to do, build a greenhouse over Nebraska?"

— Executive at a global agricultural corporation, after hearing about the solar hibernation

---

MY FIRST REACTION TO THIS executive's greenhouse idea, after we shared a short laugh, was to advise him that perhaps he should review options for planting at lower latitudes or planting different crops with more resistance to cold weather. Then later, after I hung up the phone, I realized he had the same reaction I had that day in April 2007: a feeling of hopelessness against the advancing climate change, realizing there was no way whatsoever to deflect its wrath. He seemed to sense the same stark lack of options that the projected temperature change would bring. His knee-jerk comment about Nebraska was a Freudian slip of his inner frustration over having to deal with something that, if it really happened, would be nothing less than catastrophic for him, his company, and his family. It was like the feeling you get in an auto accident. Everything appears in ultra-slow motion as you see the oncoming vehicle. You're powerless to move. All you can do is get that sick feeling in your stomach that comes when you know pain is about to strike and there is nothing you can do about it. Then, excruciating, painful reality hits you hard . . . full force . . . WHAM!

There aren't many choices for us. We are just going to have to hunker down and weather the storm. That metaphor has little comfort now, as the climate's cooling begins. We just have to use the short time we have wisely, learning as much as possible, preparing for the worst, and adapting as best as we can. That is going to be the key: adaptation. But we can adapt better if we know what is coming, if we know when it is coming, and if we have a plan to adjust to the new natural order.

We will have to downshift and retrench, just as the natural world will around us. As the Sun goes into hibernation, the plants and animals will automatically, instinctively know (most likely before we do) what is coming and how they must respond. The question here, then, is will the human race be able to adjust also? Will we have time to adjust our complex lifestyles and technologies to enable a smooth transition into decades of colder temperatures, or will we flounder in a chaotic world where no one has viable and timely solutions? Will politicians use this as another opportunity to sell us even more powerful snake oil, blaming mankind for the processes of nature?

I think the signs are abundant and crystal clear. While we are in desperate need of a national preparedness strategy to deal with the next climate change, we would be hopelessly, and unrealistically, optimistic to expect that one will be developed. Remember Hurricane Katrina and its aftermath in New Orleans? There will be no US president, no UN panel of "experts," and no congressional leaders standing up any time soon, touting the need for an international conference to identify goals and standards to address the next cold era. Will we be left stranded, waving our hands from metaphoric rooftops?

We can hope for a response to this book from the mainstream media in the coming years, but we should not expect one. Even now, after the brutal, record-setting winters of 2008–2009 and 2009–2010, and after setting new records during the winter of 2010–2011, we still see an unwavering AGW community, led by a recalcitrant presidential staff of enviro-socialists. The prospect for catastrophic earthquakes and volcanic eruptions during the current solar hibernation, fully predicted by me and other scientists, may further serve to divert attention away from cold weather preparedness.

My wife, mother, two daughters, son-in-law, and our first two grandchildren are in the same boat with every other citizen on this "blue marble" we call home. We will get no help or advice from the government, and we are not planning on it. We, who go about our daily lives struggling with layoffs, lack of jobs, moving, bills, hospitalization emergencies, caring for our parents, and helping with the grandchildren, will, as always, be on our own.

Most people will have few options when cold weather starts to affect the availability of food in the quantities we now take for granted. Americans and most people in Western nations are used to having plenty to eat. It would be a mighty struggle for most of us to adjust to less food, and we would be pretty upset if food prices doubled or tripled in one or two months' time.

So what can we do? The best plan is to have one! Here is a starting point to get you to begin considering what types of concessions you may need to make:

- Think through what the likely effects of serious food shortages will be in your home, neighborhood, town, city, state, and region.

- Are you living in the right place if the worst-case scenario develops and does so quickly?

- Do you have a sanctuary away from a major metropolitan area?

- Do you live in a region that is volcanically active or is a high risk for earthquakes, which could cause electricity, water, and communications to be out for weeks or even months?

If any of the above scenarios apply to you, do you have a plan to address them? If you think you can survive tough times with help from others, think again. They may be planning on you as their lifeline. Be the industrious ant in Aesop's famous fable and not the idle grasshopper and begin preparing now.

Friends, if there is one lasting message I can give you, it is this: become an expert in adaptability and self-reliance. For when the cold comes, jobs will be even more scarce, bread could be gone from the store shelves, and ethanol gas will be in short supply. You will find no help in newspaper articles, TV newscasts, or in the promises of politicians. You will have to prepare as best you can on your own. You need to start today . . . now!

4

The Future

"Every tomorrow has two handles. We can take hold of it with the handle of anxiety or the handle of faith."

— *Henry Ward Beecher*

DESPITE THE DIFFICULT FUTURE this book predicts for the next 30 years, it is written by someone who, believe it or not, is a dyed-in-the-wool optimist. I am also someone who gets angry when he finds out he has been deceived. I am also someone who doesn't like problems and, when they do occur, wants to get to the heart of the matter, find the root cause, and fix it so it doesn't come back. But when I do detect a problem, and especially one this all-encompassing, with such serious consequences, I am strongly driven to get it out in the open instead of keeping it quiet, or watching and waiting, ultimately allowing everyone to fend for themselves.

A friend once asked me why am I doing all this and going through the punishment that comes with spending every penny I have to keep this story alive, and in the process getting tarred by every AGW extremist and angry scientist who is peeved at someone without a PhD being so outspoken. The answer I gave was this:

Picture the scene of a group of early morning commuters standing on the platform, waiting for their regular train to take them to the big city. Two men are engaged in the usual morning conversation about jobs, the dense fog that has rolled in, and the workday ahead. The train arrives and everyone gets on — except one of the men. As the other man boards, he turns just before the doors close and the train pulls out and calls back to his friend, who mysteriously is still on the platform. "Aren't you coming into town today?" he asks. The man yells back, "No, not today; I don't think this train will ever reach town." Puzzled, the rider pays no more thought to the statement as the doors close behind him. Only ten minutes into the ride, his eyes grow wide with shock as the train leaps off the tracks, plunging down into a deep gorge because the bridge collapsed during the night. As he descends to his death amidst the screams of the other passengers, he looks back in the direction of the station and asks, "Why didn't he tell me?"

And that, I told my friend, is why I have embarked on this mission. How would you feel about me if I stayed silent and did not tell you about what I discovered on that fateful day in April 2007? It may be bad news coming during already difficult times. But with this book at least, someone is telling you the "climate bridge" is out. We have got to get off this train.

Pursuant to this personal stance, I have set a few important goals for this book that apply to all peoples and governments around the world who choose to read and heed its message. My goals are the following:

1. To alert the organizations who fight to protect our environment and preserve our flora and fauna about the potential damage we may see from the next climate change. This is so they may, where practicable, plan some form

of protection and safety net for those species that will be most affected by the bitter cold.

2. To provide advance notice to the business and investment community and government agencies charged with maintaining economic stability. This is so they can tailor economic programs to deal with the next cold era, and develop measures that can change the potentially long depression to a shorter one or, in the best case, a long recession. I doubt we can do better than a long recession. How all that will tie with current recovery efforts from the ongoing bad economy is a major issue for which there is no clear solution.

3. To give the world's agricultural organizations and businesses the opportunity to develop, test, and implement the necessary food production and storage methods for extreme cold weather that will maximize our ability to continue to feed what will be a population of over 8.2 billion in the year 2030.[1]

4. To give advance notice to the people of the world, with the strongest recommendation that everyone should develop their own survival plan — one based upon not receiving any assistance from the government at any level. Similarly, foreign governments should also base their planning on not receiving any assistance from other governments, including no corn or grain crops from the United States or Canada. This could happen soon due to potentially large-percentage losses occurring in the grain-producing areas of these countries.

5. To provide a new predictive tool, the RC theory, to help us plan for future climatic changes well before they arrive.

It is this last goal that permits me, in this chapter, to extend our vision of the future still further. This book's message is clearly one of spreading the warning of a tumultuous time ahead. But it is done so within the context of merely explaining how the Earth and Sun interact, as defined by the RC theory. This theory does not just allow us to forecast the next climate change, it also enables us to predict subsequent climate changes well into the next two centuries.

If President James Madison, in the middle of the last solar hibernation, were to have been briefed on the RC theory, he could have easily written a letter and put it in a time capsule, to be opened by the president of the United States 200 years in the future. It might have read something like this:

The White House

April 26, 1816

Dear Future Fellow President of the United States,

It has come to my attention that the Sun has remarkably predictable cycles of behavior. We appear to be in the low ebb of the cycle, bringing with it what amounts to much less heat from the Sun than normal. This behavior has had serious side effects on our weather, and it may account for the massive crop failures from cold temperatures and early snows we have had in the past few years. Many citizens of New England have had to move out by the tens of thousands to save themselves from this disastrous bad cold. Thousands less fortunate have died.

Ship captains out of Portsmouth have also reported that British and Dutch trading ships observed a major volcanic eruption two years ago, somewhere on the other side of the Earth in the Dutch East Indies. My advisers say it may account for some of our weather problems and cold, though I don't understand how a volcano on the other side of the world can make New England turn to ice in the summer.

Also, traders from out west have reported the land along the great Mississippi River has heaved and sighed like the quakes the Bible describes and that these quakes were so powerful that the very course of that mighty river was changed! We have been beset with both natural and man-made calamities during my administration. We are now trying to rebuild the Capitol after the soldiers of the king burned it to the ground a few years ago in the War of 1812, which by the grace of God, and with the help of some French warships, we won.

These have been truly difficult times for our young nation. I can only pray that our Creator will see fit to preserve our country so that there is a president of these United States in 2012 to read this, and that your times are not as grave as ours have been. If there is a United States of America in the year 2012, my advice to you is to, first, stay the course of our Constitution; second, don't let the government exceed its authority by overtaxing its citizens; and third, prepare for the cold weather and possibly some formidable earthquakes and violent volcanic activity during our Sun's cooling cycle, which is due to begin during your administration. I wish you the very best.

Sincerely,
James Madison
President of the United States

This fanciful speculation is no longer far-fetched. President Barack Obama (or his successor) may also be in the thick of the reversal of the 206-year Bicentennial Cycle and great cold climate change, as was President Madison. Similarly, the next president could also write a letter to a future president 200 years hence, alerting that person to the next major cold period. But this next letter will have the benefit of the RC theory (or similar theories)

TABLE 4-1. CLIMATE CHANGE FORECAST FOR THE NEXT 200 YEARS

Years	Climate Type	Peak/Bottom of Cycle
2020–2045	Cold Era	Bottom: 2031-2037
2070–2090	Warm Era	Peak: 2080–2085*
2110–2150	Cold Era	Bottom: 2130–2137**
2170–2235	Warm Era	Peak: 2180–2211***

* Will be less warm than the period 1990–2010.
** Will be as cold or colder than the period 2031–2037.
*** Will be less warm than the period 2080–2085.

and will be able to give a lot more information about the weather to a future world leader. It will also benefit from the study and statistics of the intervening warm and cold periods between now and then.

Based upon my research, the RC theory indicates the next climate changes will occur according to Table 4-1.

It is important to note from the table that the present 1990–2010 warm period will be the last record warm period for the next 206 years, until the next Bicentennial Cycle renews between 2180 and 2211, and even then it will be nothing compared to the warm period that just ended. The far-distant future of climate changes on Earth will then be subject to longer cycles and the more powerful influences that the larger oscillations will deliver.

This then is our climate change future, as best it can be interpreted based upon the research I have conducted as of March 2011. Future refinements in RC theory predictions are likely as additional resources are applied to further research. It is my hope that the RC theory will be fully reviewed, critiqued, and forced to stand against the best scientific scrutiny that can be mustered. If it fails the test, then so be it. I will be the first to support the talented researcher who comes up with a better concept. There is no ego to be bruised here, no research grant to be preserved, and no university tenure to be maintained.

Since this may be the first published attempt at such grand predictions, I should probably add a laundry list of caveats to this table. I will not. It is simply the best that can be done today, given the RC theory and the state of solar cycle interpretation and, to be practical, given the meager funding of the SSRC.

And with respect to funding, in early 2011, the US budget from President Barack Obama for the next fiscal year was released by the White House. By one analyst's estimate, it contains $2.6 billion devoted to global warming research in one form or another — funding the study of a climate issue that does not exist![2] Yet, at the time of this book's publication in 2014, not one government office and not one research dollar has been dedicated to the science and planning needed for the United States to be prepared for the only climate change that we can expect — a long and potentially dangerous cold climate!

I think it is entirely possible that for the next 100,000 years, the human race will not see the warmth that has been experienced by our ancestors and up to this current generation. Although I have only just begun to look at these longer and even more powerful solar cycles, my initial review, backed by many fine researchers, suggests that the next 800 years could be marked by an overall long-term and effectively permanent decline in temperatures. I would not be at all surprised to see future researchers demonstrate solar cycle behavior that concludes the Holocene warm period has finally ended and the steep drop into the next great ice age has begun in earnest.

We have comfortably rested atop the temperature curve in the Holocene warm period for about 11,000 years. Is our unusually long tenure at the top over? Comparison with other interglacial warm periods shows that our stay in this length of a warm weather cycle is indeed overextended. As such, we must derive accurate methods for predicting climate change on the order of a thousand years from now. Questions arise, such as: What will our world be like in that far, distant time? Will our learning, our technology, and our adaptability be so advanced that climate change is no longer a concern, regardless of which direction the climate is going? Will we develop the ability to live together in peace by then, to "take hold" of our species, as Henry Ward Beecher challenges us to do? Will we still be here to witness future climate changes?

Changing How the Human Species Reacts to Climate Change

We should teach such natural recurring phenomena and theories like the RC theory in elementary schools, high schools, and colleges, as we now teach about sunspots and solar flares; how tornadoes, hurricanes, and clouds form; and meteorology and climatology in general. With this theory (or a future replacement), and the outstanding work of many other researchers at their disposal, our leaders, educators, the media, and the public can enjoy a genuine sort of trust that engenders enlightened independence alongside cooperative interdependence.

I am not proposing the RC theory is the end-all of climate theories — far from it, in fact. What I am saying, though, is that it provides a host of new tools for understanding the Earth-Sun relationship, from which even greater learning can take place. If we look at our world from now on in terms of these natural cycles as a normal frame of reference, it can permanently and positively alter our perception of our place on the Earth and within this solar system. It will help bind us physically and intellectually to our cosmological environment like never before, as well as to our own planet and fellow inhabitants.

It's like riding a big roller coaster for the first time. The rapid reversals of direction can be scary because of the unknowns ahead in a highly dynamic environment. Once we realize that the ups are always followed by downs, and vice versa, and we can see the snaking of the track ahead of us, we can begin to anticipate the next reversal and brace ourselves for it. We know we won't be able to avoid it, but with each reversal we will become more accustomed to the sudden changes in gravitational tug.

So it can be in our future. If we know when and why major climate changes are likely to occur, then we can be prepared. However, if we allow others to keep us blindfolded on our natural roller coaster ride around the Sun, we as a species will be perpetually afraid of the next gut-wrenching reversal. We will continue to live in fear of the next climate change, all the while making us more susceptible to those who dredge up unnatural or politicized explanations for the situation by playing upon our fears. In such a scenario, we will remain "cold" intellectually — and when caught unprepared, we may end up cold, literally.

With the Relational Cycle theory, we can at last cast off the fear and the blindfolds. We can then see the intricate dance of the Earth, the Moon, the Sun, and the other planets for the wondrous and inspiring epic that it is. When humans someday truly understand this ballet in all its facets, I believe most will say once again, "What a thrill!"

Appendix 1: Scholarly Acknowledgments

"All truth passes through three stages. First, it is ridiculed. Second, it is violently opposed. Third, it is accepted as being self-evident."

— *Arthur Schopenhauer* (1788–1860)

HISTORY IS FULL OF accounts of people who have stepped forward to offer new ideas that ran counter to the leading theorists or dogma of their day. These new ideas, in opposition to conventional thinking, have been met with a range of responses, from silence, to ridicule, to imprisonment, and even to death at the hands of an ignorant mob, skeptics, the church, or government officials. I would like to think that we are in a different age of enlightenment and learning, and a leading-edge thinker need not fear such retribution. I regret to say, from what I have observed during the irrational, politicized period of the past global warming era, such thinking, even today, would clearly be wishful.

The disclosure that the AGW theory is essentially defunct with the coming cold period brings to mind so many past instances in science where a previous theory is eventually overcome by a greater truth — a self-evident truth.

My public disclosure of the RC theory, the research in support of my findings, and the attendant prediction of a cold era was met with a frigid reception indeed. It was, in fact, a message that no one wanted or wants to hear. My initial sense of being the only person on Earth who was aware of the coming of a global catastrophe was an incredulous, perplexing, and unnerving situation. However, in the course of the confirmation of my research and theory, I finally found, with much relief, that not only was I not alone in my conclusions about the Sun's influences on Earth's climate, but that this truth was found in abundance in the published works of numerous researchers — some of whose work appeared decades earlier than my own.

In this appendix you will see just how many researchers out there are saying we are heading for a cold climate and, more importantly, that they have been saying so for many years. As I said in Chapter 1, this book is about much more than just one researcher's theory.

The first indication during my corroboration research that I had touched on something profound and that the RC theory was on target was a paper by NASA solar physicists Dr. David Hathaway and Dr. Robert Wilson, operating out of the Marshall Space Flight Center (MSFC) in Huntsville, Alabama. In 2004, they issued a joint paper in the publication *Solar Physics* that showed the existence of what I'd termed the Centennial Cycle.[1] A review of their linchpin article helped open up a wealth of similar writings that eventually confirmed many, if not all, elements of my theory. Further research showed that still others had discovered the Centennial Cycle, as well as the Bicentennial Cycle and the other elements of the RC theory.

Here is the summary, categorized by main RC theory elements, of how much support I found by early 2007 to corroborate my research for the RC theory:

- Centennial Cycle: 18-plus researchers in over 11 papers

- Bicentennial Cycle: 17-plus researchers in over 8 papers
- Both cycles found together: 11-plus researchers in over 8 papers
- Temperature correlation with solar minimums: 9-plus researchers in over 7 papers
- Predictions of next minimum/coming cold era: 12-plus researchers in over 11 papers

Initially, I found 73 researchers in over 40 peer-reviewed papers or other publications that support all or part of the elements of the RC theory, including 16 other scientists who also predicted a coming cold era.

The tally above is by no means complete, and in fact, all are dated to 2007. After only a few weeks devoted to finding these other works in the scientific literature, I simply had to stop since it was obvious that the level of support was extensive. The total list of researchers who can lend credence to one or more elements of my theory may be 200, 300, or even 3,000 or more. I just stopped counting after the first hundred and decided to move beyond validation to pursue publishing the results of the research, publicizing my theory, and trying to alert everyone I could in the time remaining.

I did not find this list of fellow believers until after my own research was completed and I made the public announcement of the coming climate change. You can find the complete list of papers that I used during the corroborative phase of my research in the original paper on the RC theory, available at the SSRC website, SpaceAndScience.net. Below are some examples. I include myself in this list below for completeness.

Researchers who have predicted a long-term solar minimum, solar hibernation, or new climate change to a period of long-lasting cold weather based upon solar activity:

1. Dr. Habibullo I. Abdussamatov, Russian Academy of Sciences, head of space research at the Pulkovo Observatory, St. Petersburg.

   "Long-Term Variations of the Integral Radiation Flux and Possible Temperature Changes in the Solar Core," 2005, *Kinematics and Physics of Celestial Bodies*, Vol. 21, No. 6, 328–332.

   "Optimal Prediction of the Peak of the Next 11-Year Activity Cycle and the Peaks of Several Succeeding Cycles on the Basis of Long-Term Variations in the Solar Radius or Solar Constant," 2007, *Kinematics and Physics of Celestial Bodies*, Vol. 23, No. 3, 97–100.

   Comment: RIA Novosti (Russian Ministry of Communications and Mass Media), August 25, 2006: "[H]abibullo Abdussamatov said he and his colleagues had concluded that a period of global cooling similar to one seen in the late 17th century — when canals froze in the Netherlands and people had to leave their dwellings in Greenland — could start in 2012–2015 and reach its peak in 2055–2060 . . . He said he believed the future climate change would have very serious consequences and that authorities should start preparing for them today."

2. David Archibald, Summa Development Limited, Perth WA, Australia.

   "Solar Cycles 24 and 25 and Predicted Climate Response," 2006, *Energy and Environment*, Vol. 17, No. 1.

   Excerpt from paper: "Based on a solar maxima of approximately 50 for solar cycles 24 and 25, a global temperature decline of 1.5C is predicted to 2020 equating to the experience of the Dalton Minimum."

"Climate Outlook to 2030," 2007, Summa Development Limited, Perth WA, Australia.

Excerpt from paper: "The increased length of solar cycle 23 supports the view that there will be a global average temperature decline in the range of 1C to 2C for the forecast period. The projected increase of 40 ppm in atmospheric carbon dioxide to 2030 is calculated to contribute a global atmospheric temperature increase of 0.04C. The anthropogenic contribution to climate change over the forecast period will be insignificant relative to the natural cyclic variation."

3. Dr. O. G. Badalyan and Dr. V. N. Obridko, Institute of Terrestrial Magnetism, Russian Federation; Dr. J. Sykora, Astronomical Institute of the Slovak Academy of Sciences, Slovak Republic.

   "Brightness of the Coronal Green Line and Prediction for Activity Cycles 23 and 24," 2000, *Solar Physics*, Vol. 199, 421–435.

   Excerpt from paper: "A slow increase in [intensity of coronal green line in] the current cycle 23 permits us to forecast a low-Wolf-number cycle 24 with the maximum W~50 at 2010–2011."

   (Author's note: This statement translates to a coming solar hibernation.)

4. John L. Casey, Director, Space and Science Research Center, Orlando, Florida.

   "The Existence of 'Relational Cycles' of Solar Activity on a Multi-Decadal to Centennial Scale, as Significant Models of Climate Change on Earth," 2008, SSRC Research Report 1-2008: The RC Theory, SpaceAndScience.net.

   Excerpt from the research report: "As a result of the theory, it can be predicted that the next solar minimum may start within the next 3–14 years, and last 2–3 solar cycles or approximately 22–33 years. Beginning with cycle 24 but no later than cycle 25, sunspot numbers may approach a Wolf number of 50 for each of two consecutive solar cycles. It is estimated that there will be a global temperature drop on average between 1.0 and 1.5°C, if not lower, at least on the scale of the Dalton Minimum. Should the minimum begin with solar cycle 24 as forecast, the bottom of the temperature curve for this prediction is forecast for the year 2031 with widespread record cold for years on either side of 2031. A start at solar cycle 25 would extend the range of the next bottom of the solar minimum to the 2031–2044 period or more.

"Due to the predictability and accuracy afforded by the RC theory, and in the interests of the welfare of the world's citizens, the following special note is added: This forecast next solar minimum will likely be accompanied by the coldest period globally for the past 200 years and, as such, has the potential to result in worldwide agricultural, social, and economic disruption."

5. Dr. Boris Komitov, Bulgarian Academy of Sciences, Institute of Astronomy, and Dr. Vladimir Kaftan, Central Research Institute of Geodesy, Moscow.

   "The Sunspot Activity in the Last Two Millennia on the Basis of Indirect and Instrumented Indexes: Time Series Models and Their Extrapolations for the Twenty-First Century," 2004, Paper presented at the International Astronomical Union Symposium No. 223.

   Excerpt from paper: "It follows from their extrapolations for the 21st century that a super-centurial solar minimum will be occurring during the next few decades . . . It will be similar in magnitude to the Dalton Minimum, but probably longer than the last one."

"Solar Activity Variations for the Last Millennia: Will the Next Long-Period Solar Minimum be Formed?" 2003, *Geomagnetism and Aeronomy*, Vol. 43, No. 5, 553–561.

Excerpt from paper: "An analysis . . . has indicated that it is highly probable that the next long-period minimum of solar activity, which will possibly be not so deep as the Maunder and Sperer [sic] Minimums, will be formed in the 21st century." (Author's note: The Maunder and Sporer Minimums were solar hibernations that were colder than the Dalton Minimum.)

6. Dr. B. P. Bonev, Dr. Kaloyan M. Penev, Dr. Stefano Sello.

   "Long-Term Solar Variability and the Solar Cycle in the 21st Century," 2003, *The Astrophysical Journal*, Vol. 605, L81–L84.

   Excerpt from paper: "We conclude that the present epoch is at the onset of an upcoming local minimum in long-term solar variability."

7. Dr. Tim Patterson, Department of Earth Sciences, Carleton University, Canada.

   From the *Calgary Times*, May 18, 2007. "Indeed, one of the more interesting, if not alarming, statements Patterson made before the Friends of Science luncheon is satellite data that shows by the year 2020 the next solar cycle is going to be solar cycle 25 — the weakest one since the Little Ice Age (that started in the 13th century and ended around 1860), a time when people living in London, England, used to walk on a frozen Thames River and food was scarcer. Patterson: 'This should be a great strategic concern in Canada because nobody is farming north of us.' In other words, Canada — the great breadbasket of the world — just might not be able to grow grains in much of the prairies."

8. Dr. Lin Zhen-Shan and Sun Xian, Nanjing Normal University, China.

   "Multi-Scale Analysis of Global Temperature Changes and Trend of a Drop in Temperature in the Next 20 Years," 2007, Meteorology and Atmospheric Physics, 95, 115–121.

   Excerpt from paper: "The effect of greenhouse warming is deficient in counterchecking the natural cooling of global climate change in the coming 20 years. Consequently, we believe global climate changes will be in a trend of falling in the following 20 years."

9. Dr. Ken K. Schatten and W. K. Tobiska.

   Excerpt from paper presented at the 34th Solar Physics Division meeting of the American Astronomical Society, June 2003: "The surprising result of these long-range predictions is a rapid decline in solar activity, starting with cycle 24. If this trend continues, we may see the sun heading towards a 'Maunder' type of solar activity minimum — an extensive period of reduced levels of solar activity."

   (Author's note: The Maunder Minimum was centered on the year 1700 and was far colder than the Dalton Minimum of 1793–1830. Further, I regard Dr. Schatten as among the best in this field. Yet his forecast of a coming Maunder Minimum class of solar hibernation has never made it to the front pages or evening news. Should he be accurate in his prediction and the depth of cold exceed that which I have forecast, then the world is in for even more difficult times ahead than I have indicated.)

10. Dr. Y. T. Hong, H. B. Jiang, T. S. Liu, L. P. Zhou, J. Beer, H. D. Li, X. T. Leng, B. Hong, and X. G. Qin.

"Response of Climate to Solar Forcing Recorded in 6,000-Year [Isotope] O18 Time-Series of Chinese Peat Cellulose," 2000, *The Holocene* 10:1, 1–7.

Synopsis of paper: In this outstanding paper, the Chinese team of researchers observed "a striking correspondence of climate events to nearly all of the apparent solar activity changes." In showing O18 isotope measurements were high during the coldest periods, they concluded, "If the trend after AD 1950 continues . . . the next maximum of the peat O18 [and therefore cold maximum] would be expected between about AD 2000 and AD 2050."

(Author's note: This study by these Chinese scientists has a highly relevant conclusion in that climate — and hence temperatures — and solar activity tracked together "nearly all" the time for the last 6,000 years!)

11. Dr. Ian Wilson, Bob Carter, and I. A. Waite.

"Does a Spin-Orbit Coupling Between the Sun and the Jovian Planets Govern the Solar Cycle?" 2008, *Publications of the Astronomical Society of Australia* 25 (2), 85–93.

Dr. Wilson adds the additional clarification: "It supports the contention that the level of activity on the Sun will significantly diminish sometime in the next decade and remain low for about 20–30 years. On each occasion that the Sun has done this in the past, the world's mean temperature has dropped by ~1–2°C."

12. Dr. Oleg Sorokhtin, Merited Scientist of the Russia Federation and Fellow of the Russian Academy of Natural Sciences and researcher at the Oceanology Institute.

Regarding the next climate change, Dr. Sorokhtin has said: "Astrophysics know two solar cycles, of 11 and 200 years. Both are caused by changes in the radius and area of irradiating solar surface . . . Earth has passed the peak of its warmer period and a fairly cold spell will set in quite soon, by 2012. Real cold will come when solar activity reaches its minimum, by 2041, and will last for 50–60 years or even longer."

(Author's note: This forecast for duration of the solar hibernation far exceeds mine!)

13. Dr. Theodor Landscheidt (1927–2004), Schroeter Institute for Research in Cycles of Solar Activity, Canada.

    His comments from many years of research on solar climate forcing include: "Contrary to the IPCC's speculation about man-made warming as high as 5.8°C within the next hundred years, a long period of cool climate with its coldest phase around 2030 is to be expected."

    (Author's note: Dr. Landscheidt and I agree within one year, 2030 vs. 2031, on our calculations for the bottom of the next solar hibernation.)

14. Victor Manuel Velasco Herrera, researcher at the National Autonomous University of Mexico.

    His comments from his research released in August 2008: "In two years or so, there will be a small ice age that lasts for 60–80 years."

    (Author's note: Here again I disagree with use of the term "ice age," though Herrera did not define the term relative to global temperatures.)

15. Dr. Peter Harris, retired engineer, Queensland, Australia.

    From his analysis of glacial and interglacial cycles, he concludes: "We can say there is a probability of 94 percent of imminent global cooling and the beginning of the coming ice age."

16. Dr. S. Duhau and C. De Jager, University of Buenos Aires and Royal Netherlands Institute for Sea Research, respectively.

    "The forthcoming grand Minimum of Solar Activity," 2010, *Cosmology*, Vol. 8, 1983–1999.

    Comment from paper: "These [findings] lead us to conclude that solar variability is presently entering into a long Grand Minimum, this being an episode of very low solar activity, not shorter than a century. A consequence is an improvement of our earlier forecast of the strength at maximum of the present Schwabe cycle [24]. The maximum will be late (2013.5), with a sunspot number as low as 55."

Again, this is just a partial list of those who have come to the same or similar conclusions as I have on the next climate change. And only the first 13 were used in my corroborative study phase for the RC theory. Why have we not been told of their opinions? Why do we continue to hear Al Gore and the like, who have effectively no scientific background, tell us that we face an opposite climate fate from that which has been forecast by these proven researchers?

Instead of celebrities, politicians, and others pushing a flawed AGW agenda, we should be hearing from those who have been there in the trenches doing the labors of field and lab research that have given us the real, unadulterated data and solid findings to assess the future of our climate and the explanation for it. Although my RC theory has led me to make significant climate change predictions correctly, it was based upon a fairly narrow segment of the available research into the Sun's behavior and how we can measure it going back 1,200 years. This is especially true in the case of digging into the centuries of the Sun's activity before temperatures began to be measured on an official basis. This meant the use of proxies of carbon 14 (C14), oxygen 18 (O18), and beryllium 10 (Be10) isotopes taken from ice cores, benthic (deep lake and sea) mud, ancient tree rings, and so on. Then there are the endless days and nights doing sample testing, in many cases by dedicated researchers operating on paper-thin budgets in makeshift labs.

There is a large number of brilliant scientists, whose names you have never heard, who have spent their lives in basic scientific research that have then enabled people like me to connect the dots of a larger picture. This is particularly true in the field of radiocarbon dating, without which there would be no RC theory, no 206-year cycle, and no ability for me to predict the solar hibernation of the twenty-first century and the next three decades of cold weather and concurrent geophysical events.

But someone had to dig up the dots in the first place. Many dedicated researchers deserve much credit for work that directly or indirectly contributed, in some cases decades later, to my research, beyond those already mentioned in this book or listed as references in the RC theory research paper. The work of Hoyt, Schatten, and Reimar, for example, has allowed me to look back in time and then turn around and look forward into the future. My thanks to Dr. Boris Komitov for providing me with much of this list, as well as with a 1998 interview with Dr. Paul Damon, who leads the list:

> Paul Damon, Long, Sigalove, Haynes, Mitchel, Stockton, Gordon, Meko, Rubashev, A. Bonov, Anderson, Schove, Valentin Dergachev, Vladimir Chistyakov, de Vries, Stuiver, Libby, W. B. Mann, Watt, Olsson, Eddy, A. E. Douglas, Willis, Tauber, Munnich, Lingenfelter, and Ramaty.

Even this short list is missing a host of other luminaries who have contributed.

The names above are included for their contribution to the basic research that allowed me to take climate change a step further. They may not accept the RC theory, and I am unaware, among those still living, of their opinion of AGW. Nonetheless, it is their highly valued work that has garnered my respect and made them the keystone of my work. I would like to someday see a definitive scientific history of the climatic effects of the Sun on the Earth, one that recognizes the very many who have lent their scientific expertise and painstaking hours of dedication to the wealth of fundamental research that exists on the subject of climate change.

The Supporting Researchers of the SSRC

In March 2011, the Space and Science Research Center was reorganized as the Space and Science Research Corporation. With this name change came a distinguished group of scientists and researchers from around the world who wanted to lend their knowledge and distinguished credentials to my efforts in creating a greater capability to confront an ever-approaching climate menace. These individuals are, in many cases, leaders in their respective scientific fields, and are typical of those who stand up against conventional thinking, pushing the boundaries of science. They have honored me by their presence and support to the SSRC's mission of alerting the world to the next climate change:

> Dr. Boris Komitov, Dr. Ole Humlum, Dr. Dong Choi, Dr. Fumio Tsunado, Dr. Giovanni Gregori, and Dr. Natarajan Venkatanathan.

In 2007, when I announced that the world was about to undergo a dramatic shift to decades of cold weather, I was relieved to find out, even back then, that there were not only many other researchers whose work corroborated mine, but that there were also strong-minded scientists around the world willing to stand up and say, "Let's tell the full story about climate change."

Solar hibernation was discovered and publicized by many others before me, which begs the question: Why haven't we been told about this solar cycle, and why have we been given only the opposite side of the climate change story for almost 20 years?

The answer is complex. It is woven into the fabric of national and international politics, and the pursuit of environmental extremism, power, and money. It is but another sign of the greatest scientific fraud in modern times.

Appendix 2: Leadership in Climate Change Research

THE FOLLOWING EVENTS REPRESENT a record of key milestones for the Space and Science Research Corporation (SSRC) regarding its leadership on the science and planning for the next cold climate change era, based on the RC theory and the understanding that the Sun is the primary driver behind climate change.

1. **April 26–29, 2007: The final phase of an independent study into the influences of solar cycles on Earth's climate is concluded, resulting in the RC theory.** This research was conducted by John Casey, a former White House space policy advisor and space shuttle engineer, who is now director of the Space and Science Research Corporation. In addition to the theory being proposed as a model for future climate prediction, the research shows the next climate change will bring a

period of deep and long-lasting cold to the Earth, with potentially serious impacts on the world's agricultural, social, and economic systems.

2. **April 30, 2007: Initial notification is made to the White House and later to major government agencies of the coming cold era.** This is the first alert sent to the White House and the administration of President George W. Bush, warning of much colder temperatures globally as a result of the next climate change.

3. **May 2007–Present: Through a series of early press releases, John Casey begins a multiyear, comprehensive, and intensive effort to have the US government and major US and international media outlets begin covering the expected climate change to a potentially dangerous cold period.** In an active and ongoing public campaign, Casey and the later formed Space and Science Research Center become the first to alert the media and US citizens about the coming climate change.

4. **January 2, 2008: Space and Science Research Center is activated.** It thus becomes the first US research organization dedicated to the science of and planning for the next climate change to a period of long-lasting cold weather.

5. **January 2, 2008: SSRC issues its first press release, "Changes in the Sun's Surface to Bring Next Climate Change."** This release includes the use of the term "solar hibernation" to describe periods of reduced solar activity, or sunspot minimums, which result in dramatically colder climates on Earth. The SSRC website is established at www.spaceandscience.net.

6. **January 14, 2008: SSRC issues its second press release, "New Climate Change Theory Gains Influential Support."** This release includes praise from international scientists and

a strong recommendation from the past chairman of the US House of Representatives Science Committee, saying that the RC theory "should be seriously considered." At the same time, other international scientists join with the SSRC to form the first-of-its-kind research entity.

7. **January 22, 2008: The SSRC posts online SSRC Research Report 1-2008, also called the RC theory.** The peer-reviewed report is titled "The existence of 'relational cycles' of solar activity on a multi-decadal to centennial scale, as significant models of climate change on Earth." This pivotal report forms the core of research at the SSRC and includes the important prediction that "the next solar minimum will likely be accompanied by the coldest period globally for the past 200 years and, as such, has the potential to result in worldwide agricultural, social, and economic disruption." This is believed to be the first report of an independent research organization to be posted on the Internet for the specific purpose of alerting the world's citizens to the next climate change.

8. **June 2, 2008: SSRC notifies the governors of all US states of the need to prepare for the cold weather effects of the next climate change.**

9. **June 11, 2008: SSRC notifies all US senators and leaders of the House of Representatives Science Committee of the "imminent global climate change and the important effects it will have."**

10. **July 1, 2008: SSRC holds a news conference to announce, "Global Warming Has Ended — The Next Climate Change to a Pronounced Cold Era Has Begun."** The SSRC issued its third press release for 2008 concurrently. The SSRC is the first independent research organization to make such a climate change declaration, based on conclusive evidence that the

reduced solar activity and lower global temperature changes predicted by the RC theory has come to pass.

11. **September 22, 2008: SSRC sends letters to presidential and vice presidential candidates.** In Press Release SSRC 4-2008 and separate letters to the candidates Senators John McCain, Barack Obama, and Joseph Biden and Governor Sarah Palin, the SSRC asks the future US leaders to forgo climate change discussions during the remainder of the campaign and asks for rapid action to prepare the United States for the next climate change after election. The letters also warn of the potential consequences of the next climate change to a deep and long-lasting cold era, including massive grain crop losses.

12. **December 15, 2008: SSRC says Obama administration to cause worst-case climate scenario.** In a general assessment of the climate change policies of President Obama, the SSRC indicates via Press Release SSRC 5-2008 that the new administration's climate policies will result in a combined effect of misspent funds, lost time, and lack of preparedness for the next climate change. This will result in a cold weather era more destructive than already forecast.

13. **January 3, 2009: A study of websites shows the SSRC is the most referenced site for the next climate change.** A random search of generic key words "climate change cold era" shows the SSRC is the most often quoted resource on the World Wide Web for information on the next cold climate period.

14. **January 8, 2009: Presidential appointees Dr. John Holdren (Obama administration science advisor) and Dr. Jane Lubchenco (NOAA administrator) are notified of the end of global warming.** In a letter and Press Release SSRC 1-2009, top administration officials are told of the status of

Earth's climate and that "global warming is over, a new cold climate has arrived."

15. **March 2, 2009: SSRC sends final request to President Obama to reverse course of climate change policies.** Also announced in SSRC Press Release 2-2009, the SSRC makes a last appeal to the White House to change climate policy direction or else create risks for the people and economy.

16. **March 19, 2009: The SSRC notifies all state attorneys general and US attorney general Eric Holder that the Securities and Exchange Commission (SEC) should review climate change policies to protect investors.** In private letters to all attorneys general in the United States, the SSRC advises them of the potential illegality of climate change instruments and provides copies of the letter sent to SEC chairman Mary Schapiro.

17. **June 15, 2009: SSRC requests that the Securities and Exchange Commission protect investors from climate change policies.** The SSRC requests the SEC to halt all carbon trading and make investors aware that global warming has ended and climate change instruments proposed by the Obama administration may be "worthless securities."

18. **June 17, 2009: The SSRC releases its detailed forecast for the extent and schedule of the coming cold climate era.** In order to address a common question about the next climate change, Director John Casey provides the years and depth of cold temperatures estimated for the new cold era up to the bottom of the cold period in the 2030s. The most definitive temperature profile ever published for the next climate change, it also discusses the lack of preparedness in the United States because of the climate change policies of President Obama's administration.

19. **July 13, 2009: The director of the SSRC, John Casey, in Press Release 5-2009, calls for the firing of the president's science advisor and NOAA administrator.** This call is prompted by the report "Global Climate Change Impacts in the United States" issued to Congress and the public. This report is described by Casey as "a piece of blatant, politically motivated bad science and pure propaganda intended to reinforce the 'big lie' that global warming is still a threat to the planet."

20. **March 1, 2010: The SSRC issues its second Research Report 1-2010 (Preliminary) and associated Press Release SSRC 1-2010. The report is titled "Correlation of Solar Activity Minimums and Large Magnitude Geophysical Events."** This research report provides substantial evidence of the likelihood of major, possibly historic volcanic eruptions and earthquakes to occur during the ongoing solar hibernation. In this report, a high percentage of probability is established for major geophysical events based on 400 years of the largest earthquakes in the United States and volcanic eruptions worldwide. The press release is titled "Sun's Activity Linked to Largest Earthquakes and Volcanoes."

21. **May 10, 2010: The SSRC issues Press Release 2-2010 titled "Food and Ethanol Shortages Imminent as Earth Enters New Cold Climate Era."** This release warns, for the first time in the United States, of near-term potential for serious agricultural losses from the impending cold climate. It includes notification of Secretary of Agriculture Tom Vilsack and FBI Director Robert Mueller of the coming potential for social and economic disruption caused by projected crop/food damage.

22. **January 25, 2011: Solar hibernation, predicted by John Casey, is confirmed by NASA data.** In a special Press Release, 1-2011, the SSRC announces that at last, the four-year effort of its director, John Casey, to convince the scientific

community, the US government, the media, and the public that a solar hibernation is coming has now been substantiated. The virtually automatic 20- to 30-year global cold era that always follows such hibernations also predicted by Mr. Casey is also validated.

23. **February 4, 2011: The SSRC issues Press Release 2-2011, showing that the next phase of global cooling has begun as marked by a record drop in Earth's ocean temperatures.** Relying on NASA and NOAA satellite measurements and processed by other government-contracted organizations, the SSRC press release makes note of how this new record steep drop in ocean temperatures supports its May 10, 2010, forecast for a record global temperature reduction by December 2012.

24. **February 7, 2011: The SSRC reorganizes and is joined by leading scientists, as discussed in Press Release 3-2011.** The SSRC is renamed the Space and Science Research Corporation and receives the support of some of the world's leading scientists as they join the list of supporting researchers at the SSRC.

25. **March 14, 2011: The SSRC issues Press Release 4-2011, warning that there will be more and larger earthquakes than that which struck offshore near Sendai, Japan, on March 11, 2011.** The SSRC reinforces its March 2010 prediction for historic earthquakes and volcanic predictions because of the advent of a solar hibernation.

Appendix 3: Press Releases: Global Warming Has Ended; The Next Climate Change Has Begun

"There is not a truth I fear, or would wish unknown to the whole world."
— *Thomas Jefferson,* Letter to Henry Lee, May 15, 1826

THE MATTER OF A new theory proposing an advancing climate change, with its potentially harmful global cooling, in direct opposition to what the media and US government are telling you, demands a strong set of evidence.

This is especially the case since it predicts more immediate and substantially more dangerous effects in the near term than the opposite theory of anthropogenic global warming (AGW). Effects from AGW aren't supposed to produce major adverse impacts until the year 2100, or at least so we have been told.

In this appendix, I will provide for your consideration 33 compelling reasons for you to accept that global warming has in fact ended and the new cold climate has begun.

The necessary proof of the arrival of a new colder climate may be quickly demonstrated by global temperature charts showing that Earth has in fact started a long-term cold weather trend. You will certainly see that evidence in this appendix.

However, the substantial collateral distress I have predicted from this new climate mandates a stronger, more detailed coverage of the subject. Of equal importance is the requirement to show that a solar hibernation has started. It is the linchpin to a forecast of decades of much colder weather.

It all began with this press release . . .

Press Release

April 30, 2007

A Period of Global Cooling Has Begun

Research into a planned book on natural catastrophes has resulted in startling conclusions on the issue of global warming, GW.

According to Mr. John L. Casey, the book's author, the book was not originally intended to look at the subject of global warming. Mr. Casey explained that he was deep in the research for the portion of the book dealing with solar flares and other destructive effects of the Sun's activity, when he found out what could have major ramifications for the global warming debate and future climate forecasting.

"What I found was totally unexpected," said Casey. "I was not even going to come close to the highly contentious subject of global warming but instead just concentrate on Sun-related dangers facing people on Earth. It was amazing to see the correlations between the Sun's energy cycles and the global temperature history. The more I looked at the data, the more obvious it became that there were recurring solar cycles that were not just contributing to the heating and cooling cycles of the Earth, rather they were the most dominant and highly predictable cycles covering at least the last 1,100 years."

He has termed these cycles "relational cycles," since the two main cycles occur in periods of time that cover one or two human lifetimes, within which one can experience or relate directly to these cycles' actions on the weather. Once the initial discovery was confirmed by comparison with several sets of data from differing sources, he began an even more extensive examination of available research and data on the subject, well beyond the original scope of the book.

"Most of the past century and until now, we have been under the direct influence of the very powerful 207-year cycle," revealed Casey. "It is the primary reason for the excessive heating we have experienced on Earth over the past 30 years. If the sunspot count reaches the 70s or 80s or less in the year 2012, then that will definitively signal the end of the 207-year cycle. I was stunned by the almost 100 percent correlation between solar activity lows and global and US low temperatures going back hundreds of years. I don't recall ever seeing this strong a relationship in natural systems before."

As to when his theory of relational cycles says that the next big changes will take place, Casey advised, "Regardless whether the 2012 solar cycle or following cycle is the changeover point, I believe there is little doubt that we are about to enter a prolonged period of rapid cooldown. Preliminary data and early 2006 to 2007 comparisons suggest it has already started."

In view of the fickle business of forecasting weather, much less global weather patterns, Casey added a cautionary note. "The weather can turn on a dime, and this spring has already shown a sometimes dangerous variety of weather. We may see a few more high temperature records being set this year or next, for example. Nonetheless, the end of the current 207-year cycle should be obvious to all by 2012. I am encouraged to see what may be early signs of the cycle changeover. Sunspot counts have been dropping rapidly, we've seen a recent record cold temperature in Alaska, and the preliminary signs of a return to more normal seasonal temperatures are coming in. Though March was another record setter for high temperatures, January and February were normal and April may be normal to slightly cooler than normal. Comparisons from 90-day periods from 2006 to 2007 are displaying significant temperature reductions. We must go through normal temperatures again, in the transition from the record heat of the past to reach the coming cold period ahead."

Asked why he was able to uncover his theory of relational cycles in view of his lack of experience in climatology, Mr. Casey lamented, "Someone should have pulled this together much sooner. From the news reports and TV coverage and comments I have already gotten as I start getting the word out, it appears that others may have been reluctant to go out on a limb because of the 'climate' of the climate debate. Many very talented field researchers have been doing outstanding work in this area for 15 or 20 years. I am quite comfortable looking at complex scientific challenges, and the field of astronomy and physics is near and dear to me. My education and space program background speaks to that. Without any assistance other than access to raw data sets and charts, I was able to discern the underlying trends, develop my theory, and then confirm my findings with some of the best and brightest in this and other countries. Someone else should have announced this before me."

As for whether he is ready to be criticized for his work, Casey was resolute, saying, "I expect criticism because this revelation flies in the face of conventional thinking. Being in that position is not new to me. On the other hand, I have a powerful ally on my side, the Sun. The Sun has been making substantial hot to cold cycle changes in an almost clock-like fashion for the past 1,100 or more years, maybe much longer. More research into relational cycles and larger time frame cycles is needed. Besides, whereas the main body of climatologists and the UN's latest reports say that that we need to wait for the year 2050 or later to see if their climate models' predictions are correct, my theory will be validated just around the corner in 2012, maybe much sooner. And the signs that my theory is correct are already appearing.

"Based on other research I am doing for the book, we have much bigger problems on the horizon. If I'm correct, the global warming rancor will be over soon. If not, then that means that one of the Sun's fundamental cycles of heating and cooling the Earth has changed. Should that happen, we are in for real trouble."

And so the snowball was started downhill — the first public press release had gone out.

There is a reassuring aspect to the scientific method that should give us all comfort. Even though Galileo was placed under house arrest by the Catholic Church for his scientific theories of the Sun being at the center of the solar system instead of the Earth, and for centuries, the Earth was widely held to be flat, science eventually prevails over dogma, ignorance, and politics. In the world of theories, new ones routinely replace older ones as we become day by day, digit by digit, synapse by synapse, a little more knowledgeable than we were the day before. Sometimes, this occurs in a slow, methodical progression. Other times, it takes place with a great leap forward. Regardless of how it arrives, once a better theory comes along, it eventually replaces the older one. It is the logical course of human reasoning and a foundational beauty of the pursuit of science.

Historically, though, this process has always been delayed or corrupted by those in power, or those who want to be, those whose influence would be reduced, or those who perceive their reputation may be tainted or their own theories invalidated by a new theory. In my own case, there have been times in the past when I also held beliefs that were so strong that I would have considered them eternal. Yet, when new revealing research came along and showed my position was wrong, I eventually (in some cases, grudgingly) changed my beliefs. I am just as convinced that there are many dedicated scientists and supporters of the AGW theory who are totally assured they are still on the right path. I am likewise convinced that, upon review of my new theory and especially the advent of the tangible and validated global climate changes and changes in the Sun consistent with my theory, those with objective, open minds will also adopt "a better theory."

There is ultimate salvation for a civilized society. It comes from free and open debate, illuminated by the light of truth.

Let there be light!

Now, let us look at the 33 reasons I mentioned that demonstrate why global warming has ended and a new cold climate has begun:

REASON 1:

A formal declaration of the end of global warming and the start of the next climate change has been made by a competent authority.

After my first press release of April 30, 2007, the forecast for a change to a new cold weather climate was viewed by most as, well, just another climate theory. In fact, it was seen with even less credulity since it was proposed by someone essentially unknown in the professional climate science community, a person without any past record of university research and not one published paper in any scientific journal. Regardless of my space program and engineering background, not having a PhD after my name or any background in climate science was (and still is) an obstacle to overcome. For some, if one doesn't have the right title, then their opinion just doesn't measure up, regardless of whether the evidence is rock solid, the research is echoed by other professionals in the field, and predictions have come to pass exactly as forecast.

In 2007, I was so certain of my research and the reliability of my findings and the forecast of the start of the next solar hibernation within three years that I sent out the first announcement before I did the corroborative study to see whether any other scientists had come to the same conclusion!

In retrospect, I could have taken the more cautious approach and done my corroborative research before releasing my findings. Clearly, it was a big risk, but I had confidence in my findings and conclusions. Time has shown I made the right call. Within one year, however, something crucial to the RC theory and its predictions took place: the theory became validated. According to available signs, global warming had ended and the momentous changes in the Sun forecast by me via the RC theory began to appear. The RC theory was no longer just a theory, it was reality! It was a climate change prediction tool that actually worked!

The Declaration of the End of Global Warming and the Start of a New Cold Climate

By the middle of 2008 and for several reasons, it was obvious to me that it was time to make an unequivocal, public, and formal announcement of this advancing new climate period. But a formal announcement of something of this significance demands a solid footing and basis that is unquestionable. There had to be several key elements in place to warrant the nature, content, and relevance of this kind of critical climate change statement. I set the following preconditions before finally making the declaration:

1. **A significant short-term drop in global temperatures had to occur.** This happened between 2007 and 2008, bringing Earth's temperature back to levels of the 1980s (if only briefly).
2. **A sustained long-term trend of the start of colder temperatures had to be demonstrated.** This trend was demonstrated by global temperature charts of independent monitoring stations by mid-2008.
3. **The forecasted simultaneous start of a solar hibernation was required.** The hibernation began as I predicted.
4. **There had to be an absence of any other science organization ready to make a declaration.** There was no other organization prepared to release this evidence, as the SSRC was.
5. **There had to be a scientific theory — an explanation that accounted for the change in climate.** The RC theory met this requirement.
6. **There had to be an organization with the recognized expertise to lend credibility to the declaration.** There was — the SSRC. NASA, NOAA, and the United Nations Intergovernmental Panel on Climate Change (IPCC) all missed the end of global warming, the drop in Earth's temperatures, the start of the solar hibernation, and the change of climates. The SSRC correctly predicted all these major climate science events in advance. This record, along with acknowledgment of SSRC research by other scientists, gave the SSRC the credibility it needed.

In addition to the above preconditions, two other factors drove the decision to declare an end to global warming.

1. **There was a critical need to change the debate from preparing for global warming to preparing for the next**

cold climate period. Precious time and resources were being spent on the debate over global warming — a climate that no longer existed.

2. **A series of important events in support of man-made climate change demanded a corresponding opposite and decisive statement appropriate to the new climate reality.** The years 2007 and 2008 were filled with a maximum effort by the media, world governments, and the Nobel Prize committee to recognize man-made global warming as a threat. Someone outside the greenhouse gas debate had to stand up and give a new compelling reason that we should put an end to this process.

On July 1, 2008, in Orlando, Florida, I put together a news conference with two local TV stations. Here is the critical statement that came from that news conference:

> "After an exhaustive review of a substantial body of climate research, and in conjunction with the obvious and compelling new evidence that exists, it is time that the world community acknowledges that the Earth has begun its next climate change. In an opinion echoed by many scientists around the world, the Space and Science Research Center (SSRC) today declares that the world's climate warming of the past decades has now come to an end. A new climate era has already started that is bringing predominantly colder global temperatures for many years into the future. In some years this new climate will create dangerously cold weather with significant ill effects worldwide. Global warming is over — a new cold climate has begun."

Immediately after the news conference ended, a corresponding press release was posted on the Internet (SSRC Press Release 3-2008) and e-mailed to many leaders in Washington and the mainstream media, including financial magazines and services. Soon after, three important outcomes were observed:

1. The Chicago Carbon Exchange began a free fall of the valuations in its credit trading. Within a few months, the greatest part of the worth of this market was gone. Two years later, its much hyped "$10 trillion" operation had all but disappeared. According to an extensive article by Patrick Henningsen on September 6, 2010 (themarketoracle.com), carbon credits were $7.40 per ton in June 2008. The declaration of the end of global warming (i.e., there was no need for carbon credits) was July 1, 2008. As of his article, carbon credits were trading at ten cents per ton! It is possible that billions of dollars may have been lost in this clearly fraudulent operation based on trading in what were always worthless securities to fix a climate problem that did not exist, supported by an unproven theory.

2. The AGW movement, including the climate czars in the White House, quickly and quietly stopped the widespread use of the term "global warming" and replaced it with a more nebulous yet potentially longer-lived term: "climate change." With this subtle adjustment, they could now claim any global change in climate, whether warming or cooling, could somehow be pinned on mankind.

3. Internet talk about a new cold climate spiked upward and has been rising since. By the winter of 2009–2010, a little over a year later, there was even talk about a "new ice age" beginning. This is not a term I agree with, but it signified the start of a change in mind-sets nonetheless. A Google search conducted in January 2009 on the generic words "climate change cold era" showed that the SSRC was the most widely quoted organization on the web regarding this new climate era.

It may be difficult to specifically tie the SSRC declaration of the end of global warming to these three significant events in the war for truth in climate change. The 2007–2008 temperature drop was a big enough wake-up call by itself. Those behind the scenes are hardly going to step forward to tell why and when they made decisions to shift to the new climate reality. Yet, the timing is at least highly suspicious.

In retrospect, this small news conference I slapped together in July 2008 may be the most important event I ever attended. It is often difficult to see through the fog of present actions and confusing signs to understand the future. At the time, I guess I was simply following an old adage of the frontiersman and folklore hero Davy Crockett: "Make sure you're right. Then go ahead."

REASON 2:

A major short-term drop in Earth's temperature took place between 2007 and 2008.

To the great consternation of the AGW community and the IPCC, Earth's temperatures have not been rising steadily as predicted. This thumb in the IPCC's eye by Mother Earth took place on the first go-around between 2007 and 2008.

Though most news outlets ignored the historic drop in global temperatures during 2007 and 2008, it was not lost on the web or the climate science community and certainly not among those who opposed the fraud of AGW. The reporting of the 2007–2008 drop in temperatures was also reported widely on the Internet among those sites that have a record of climate change skepticism and objective analysis of the whole subject area.

There are several temperature-monitoring institutions reporting on global temperatures. The most widely followed are the Hadley Centre for Climate Prediction and Research and the University of East Anglia's Climate Research Unit (CRU), both in the UK; the University of Alabama, Huntsville (UAH); Remote Sensing Systems (RSS), a company under contract to NOAA/NASA; the Goddard Institute for Space Studies (GISS), a NASA lab; and the National Climate Data Center (NCDC), a NOAA unit. Hadley and CRU have cooperated in development of a global temperatures series, called the HADCRUT. They have also served as one of the primary sources of data for the IPCC.

The Hadley Centre, UAH, and RSS all published their measurements of the significant drop in Earth's temperatures in early 2008. Each established on their own that Earth had gone through a historic drop in average temperatures between January 2007 and mid-2008.[1, 2, 3] UAH saw a drop of 0.588°C between January 2007 and January 2008. The Hadley Centre, over the same period, reported a drop of 0.595°C. Between January 2007 and May 2008, RSS reported a drop of 0.643°C. These temperature reductions don't seem like much, but they brought global temperatures back in line to where they were during the 1980s.

In other words, the bulk of the global warming we had heard so much about for the previous 20 years had been nullified, if only for a year and a half. I discuss in this book the fault others make in using short-term temperature corrections as indicators of a trend, and I have done my best to avoid this trap. Still, this major global temperature reduction from 2007 to mid-2008 was important for its depth and duration, and for the utter silence it created within the media and AGW community.

REASON 3:

Another record drop in global temperatures was predicted between May 2010 and December 2012.

For most of 2009 and 2010, I received numerous requests from friends and website followers who wanted to know when the first cold damage to crops could be seen. Some actually wanted my recommendations on where they could move to avoid the next 30 years of crop-depleted cold climate. In April and May 2010, I finally relented and did more detailed analysis of when the next major cold weather drop would occur and how soon that might translate into noticeable crop damage. I predicted that this would take place between May 2010 and December 2012.

The forecast record cold that occurred during that period was another convincing sign of the new cold climate era, with the added threat of the first signs of the danger it poses to the world's agricultural systems. This event will provide a shocking realization to all of Earth's people and their governments that global warming has ended, a new cold era has begun, and the Sun — not mankind's greenhouse gas emissions — determines climate changes.

This predicted historic drop will bring the first meaningful damage to US crops sufficient to cause real worry by the public. With the forecast of crop damage, we will also see the concurrent loss of corn crops that have created the lunacy of using food for fuel. Surely history will look back on the United States during this period and proclaim it as one of the darkest moments in the course of human civilization. The following press release was put out on May 10, 2010:

Press Release SSRC 2-2010 (edited)

Food and Ethanol Shortages Imminent as Earth Enters New Cold Climate Era

Monday, May 10, 2010

11:30 AM

The Space and Science Research Center (SSRC), the leading independent research organization in the United States on the subject of the next climate change, issues today the following warning of imminent crop damage expected to produce food and ethanol shortages for the United States and Canada:

Over the next 30 months, global temperatures are expected to make another dramatic drop even greater than that seen during the 2007–2008 period. As Earth's current El Niño dissipates, the planet will return to the long-term temperature decline brought on by the sun's historic reduction in output, the ongoing "solar hibernation." In follow-up to the specific global temperature forecast posted in SSRC Press Release 4-2009, the SSRC advises that in order to return to the long-term decline slope from the current El Niño-induced high temperatures, a significant global cold weather redirection must occur. According to SSRC Director John Casey, "The Earth typically makes adjustments in major temperature spikes within two to three years. In this case as we cool down from El Niño, we are dealing with the combined effects of this planetary thermodynamic normalization and the influence of the more powerful underlying global temperature downturn brought on by the solar hibernation. Both forces will present the first opportunity since the period of Sun-caused global warming period ended to witness obvious harmful agricultural impacts of the new cold climate. Analysis shows that food- and crop-derived fuel will, for the first time, become threatened in the next two and a half years. Though the SSRC does not get involved with short-term weather prediction, it would not be unusual to see these ill effects this year much less within the next 30 months."

The SSRC further adds that the severity of this projected near-term decline may be on the order of 0.9°C to 1.1°C from present levels. Surprising cold weather fronts will adversely impact all northern grain crops, including of course wheat and the corn used in ethanol for automotive fuel.

In pointing out the importance and reliability of this new temperature forecast and its effects on North American crops, Director Casey adds, "The SSRC has been the only US independent research organization to correctly predict in advance three of the most important events in all of climate science history. We accurately announced beforehand, the end of global warming, a long-term drop in Earth's temperatures, and most importantly the advent of a historic drop in the Sun's output, a solar hibernation. The US government's leading science organizations, NASA and NOAA, have completely missed all three, as of course have United Nations climate change experts. It is only because of the amount of expected criticism we received because of our strong opposition to the Obama administration's climate change policies and our declaration of the end of global warming, that the SSRC is not more fully accepted for its leadership role in climate change forecasting. The facts and reliability surrounding our well-publicized predictions however stand as testament to the SSRC's proven ability to understand the nature of global climate change. In view of the importance of this new forecast, I have notified the Secretary of Agriculture to take immediate actions to prepare the nation's agricultural industry for the coming crop damage."

The SSRC places only one caveat on this forecast. Casey elaborates: "Only a stronger solar cycle with a period longer than the 206-year cycle can cause us to alter our projections. Although more research is needed in this area, none have yet shown themselves. The present hibernation is proceeding in almost lockstep as the last one, which occurred from 1793 to 1830. If it continues on present course, while the cold weather impacts on food and fuel announced today are certainly important, they do not compare with what is to follow later. At the bottom of the cold cycle of this hibernation in the late 2020s and 2030s there will likely be years with devastating to total crop losses in the Canadian and northern US grain regions."

This press release predicted a new record drop in global temperatures by November-December 2012. It also made it clear that this solar minimum was proceeding just like the last one.

In my now characteristic manner, I issued this release about record cold weather smack dab in the middle of NOAA's proclamations, along with that of Dr. Hansen at NASA's Goddard Center, that 2010 was going to be a record hot year — perhaps the warmest on record. Once again, my statement, in direct opposition to conventional thinking, did not make it to the evening news or the front page of any newspapers. Guess what happened?

REASON 4:

A major decline in global temperatures has started, reinforcing the SSRC prediction for a historic temperature reduction by December 2012.

No sooner than I made my prediction in May 2010 — the Sun obliged with the termination of the warming that had marked the start of 2010. By mid-2010, the heat was waning and the next short-term global temperature drop had started. Winter came early in November and began to pound the United States with record cold and snow in a series of unprecedented storms. In early February 2011, I issued another press release showing the support for my May 2010 prediction in another attempt to call attention to the cold that was quickly approaching us:

Press Release 2-2011 (edited)

Global Cooling Begins and Global Warming Ends with Record Drop in Temperatures

Friday, February 4, 2011

5:00 PM

The Space and Science Research Center (SSRC) announces today that the most recent global temperature data through January 31, 2011, using NASA and NOAA weather satellites, supports the previous forecast from the SSRC that a historic drop in global temperatures is underway and that the previously predicted climate change to one of a long and deep global cooling era has begun.

SSRC Director John L. Casey explains, "Based on the data from the AMSR-E instrument on board the NASA Aqua satellite, sea surface temperatures just posted this week showed their steepest decline since the satellite was made operational in 2002. This major drop from the warm temperature levels seen in 2010 is also echoed by a dramatic decline in atmospheric temperatures in the lower troposphere, where we live, with the data coming from NOAA satellites. At present rates of descent, both ocean and atmospheric temperatures are likely to soon surpass the temperature lows set in the 2007–2008 period. Even with a small correction that is usually seen after such a rapid drop, there is no doubt that the Earth is entering a prolonged global cooling period and will soon set another record drop in temperatures by the November-December 2012 time frame, as was forecast in the SSRC press release from May 10, 2010."

As to the long-term implications of this significant drop in global temperatures, Director Casey clarifies by adding, "While we always see a reduction from a previous El Niño high, this time the decline is different, very different. What is happening now is the effect of the natural La Niña cooling is being overpowered and accelerated by a once every 206-year solar cycle that has entered its cold phase. In 2007, after discovering this cycle, I was the first to announce to the White House, Congress, and the mainstream media that this cycle would produce a 'solar hibernation,' a major reduction in the output of the Sun which in turn would bring a new climate change to a cold era lasting 20–30 years. This hibernation, also called a grand minimum, was recently verified by NASA data using sunspot measurements and was announced in another SSRC press release January 25th of this year. In quick succession, here in early 2011 we have seen two of the strongest possible validations of the global cooling phase of the 206-year cycle and the Relational Cycle theory of climate change which I developed to account for the pattern of alternating cold and warm periods that we have seen for over 200 years now. Although we will continue to see highly variable weather, the punishing winters the world has seen the past few years, including the ongoing record-setting winter of 2010–2011, are just a sample of what is to come.

“Though the conclusions of my research and that of many others around the world has shown a new and potentially dangerous cold weather period is coming, the recent NASA data about the Sun going into hibernation and this week's global temperature figures have provided critical evidence for our leaders and the public to finally see that the next cold climate era is here.

“It is also important to recognize that there has been no effective growth in Earth's temperatures for 12 years now and according to my calculations, the statistical peak of the long-term curve of the past Sun-caused global warming was probably between 2005 and 2007. Global temperatures have suddenly returned to the same level they were in 1980 and are expected to drop much further. Given the momentum of the solar hibernation, it is now unlikely that our generation or the next one will return to the level of global warming that we have just passed through. Again, global warming has ended. It was always caused by the Sun and not mankind. The global cooling era has begun.

“The SSRC has a track record for accuracy in climate predictions that is among the best. It remains the only independent research organization in the United States that has been consistently warning the US government, the media, and the public that this new cold weather is upon us and that we need our people to prepare. As stated many times before, this solar hibernation will bring the worst cold in over 200 years and will likely cause substantial damage to the world's agricultural systems. Here at the SSRC, we will continue to post these releases with new updates so our citizens are well informed."

The satellite temperature data is available through several NASA and NOAA sources, including Remote Sensing Systems (RSS) out of Santa Rosa, California (www.remss.com), with both sea and atmospheric temperature charts available from the University of Alabama, Huntsville (UAH), via the website of UAH's Dr. Roy Spencer (www.drroyspencer.com).

By the time this book is published, we will have seen even more global temperature data and possibly a start of quick warm reversal that typically comes with these record steep reductions. The long-term picture will not change, however, as the solar hibernation works its will on the Earth.

REASON 5:

A long-term global temperature trend toward a colder climate has been established in Earth's atmosphere.

Data from several sources documents my previously predicted long-term decline in Earth's temperatures. The charts in Figures A3-1 and A3-2 show where the Earth has been and where it is going when it comes to its future temperature profile.

This larger picture of Earth's temperatures from www.climate4you.com gives us a longer-range perspective on the temperature variations we have experienced for the past century. In the middle of the gently waving temperature curve, we can easily see the 40 years of

100 YEAR TREND ANALYSIS

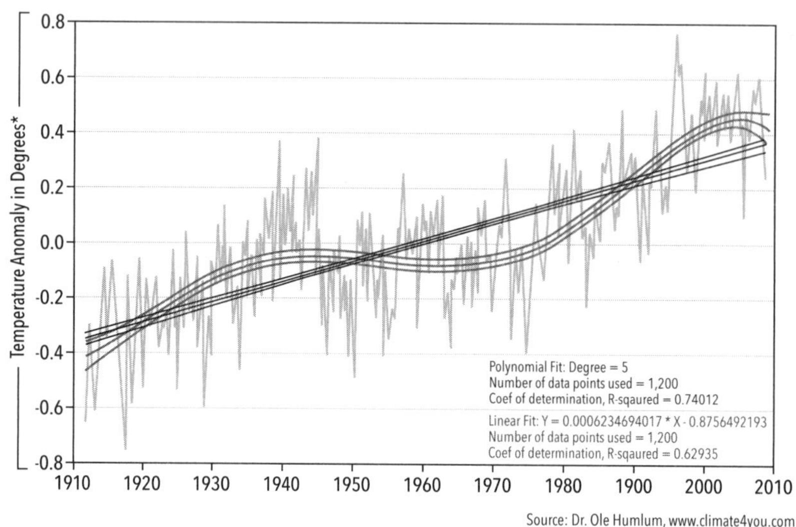

Figure A3-1. Global temperatures for the last hundred years. This excellent chart provides more convincing evidence, up to February 2011, not only that global warming has stopped, but also that the long-term decline in Earth's temperatures predicted by the RC theory has begun. *Reference period: 1961–1990.

temperature decline from roughly 1940 to 1980. The overall trend, of course, has been one of growth, as shown by the straight line linear trend. Still, at the top right of the chart we can see something very important has happened: the curve has begun to reverse direction once again, and steeply so. This area at the top of the curve represents 20 years of the man-made "global warming" era, during which we have come under tremendous pressure from politicians, environmental groups, and the mainstream media to accept the man-made global warming theory. A closer examination of the last two or three decades, which you can see in Figures A3-2 and A3-3, reveals a totally different understanding from what we have been led to believe.

Climate4you.com operates one of the best sites on the web, providing numerous charts and links to global climate change

30 YEAR TREND ANALYSIS

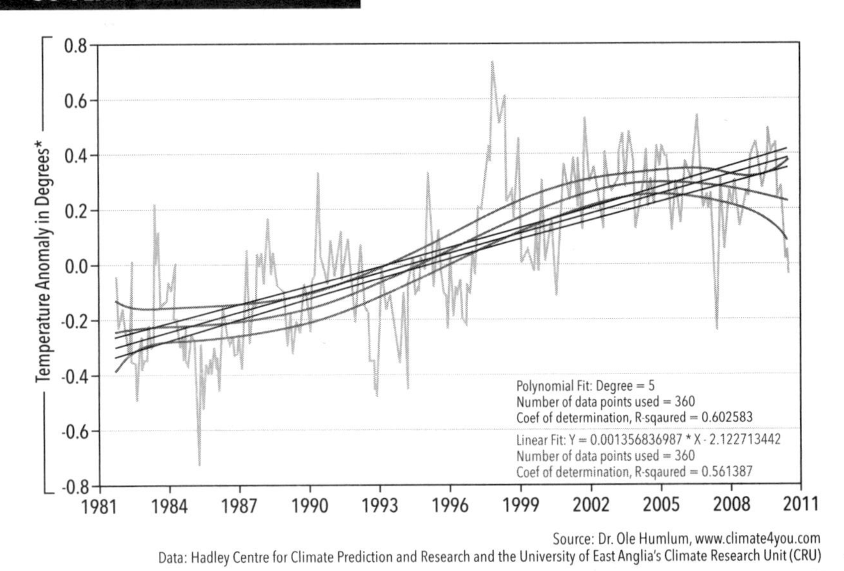

Figure A3-2. Global temperatures for the last 30 years. This chart from the excellent website of www.climate4you.com by Dr. Humlum displays the now evident long-term change in Earth's temperature. *Reference period: 1961–1990.

and temperature sources. I heartily recommend it to the objective reader and student of climate science. This site of Dr. Ole Humlum, who is a professor in geology and an expert on glaciers at the University of Oslo, is both balanced and deep in research material and authoritative sources.

Here in this 30-year span of time, we can see signs of the next climate as it begins to appear close-up. My previously estimated end of global warming between 2005 and 2007 is visible at the top of the sloping temperature curve, where it starts to flatten out. The straight line still shows an overall trend that the AGW movement has been relying on to claim Earth's temperatures can only go up. As is the case with any straight line trend, it only shows accurately where you have been. In the case of naturally occurring and variable cycles, especially long ones of 100 to 200 years, it does not show where you will be going.

REASON 6:

There was no growth in Earth's temperatures for 13 years.

You can prove this assertion easily yourself. In the 30-year chart above, take a straight edge or ruler and lay it across anywhere in the 1997–1999 time frame and extend it to where the chart ends in February 2011. The new straight line trend is distinctly flat or declining for 13 years. This point is clearly depicted in Figure A3-3, a chart of the temperatures for the 10-year period from 2001 to 2011.

Again, during this period of a dozen years or so and continuing until today, we have been subjected to the most aggressive propaganda ever to try to convince the world that CO_2 will cause the Earth to warm dangerously, that temperatures are on an

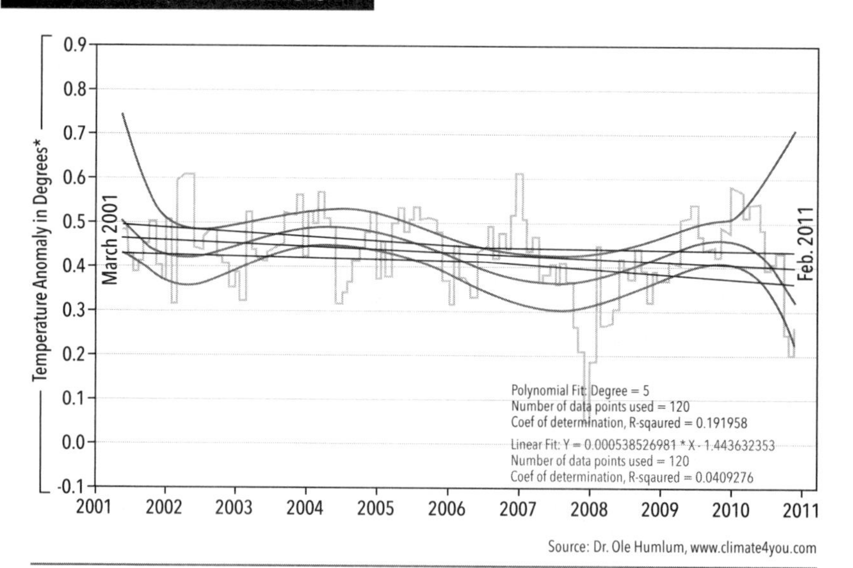

Figure A3-3. Global temperatures for 2001–2011. This chart shows that we have had at least ten years where the world's temperature through February 2011 had begun to decline. Global warming has stopped without a doubt. The line spikes at the beginning and end of the top curve are the result of how the chart is constructed and not a temperature measurement. Extending this back to 1998 shows 13 years with no global warming and instead a continuous drop in temperatures. *Reference period: 1961–1990.

uncontrollable path of growth, and that mankind's greenhouse gasses will overheat the planet, causing melting glaciers to flood the world's major cities. The straight line curve in the chart above also shows where we have been — on a declining path of global temperatures. The curved line shows where we are going — down. Global warming as a continuous process of ever-increasing global temperatures has ended. It ended years ago.

In the peak year of global temperatures, 1998, AGW advocates reveled in the moment, saying it proved man-made global warming. It is ironic that, in fact, 1998 was the beginning of the end of global warming. Because they did not understand what was causing the warming, they certainly could not see why it was about to end.

REASON 7:

We have been misled by climate change alarmism for over a hundred years because of a narrow view of when climate change occurs and why.

A review of climate change history and how the media and the scientific community have reacted to it over the past century confirms the tendency to use a small span of time to make erroneous conclusions about what Earth's climate will do in the future, much less its underlying causes for change. According to the RC theory, whose correlation of global temperatures (climate) to sunspot records is accurate to over 90 percent, major climate changes can only be discerned by an understanding of solar cycles 100 to 200 years or longer in duration. Such a "blinders on" view of global climate using smaller periods of time, including the recent 20-year span of warming, has often led us astray, as the following table illustrates.

Let's look back to June 24, 1974, to see an example of how one of these climate alarms was sounded. Here is a quote from a *Time* magazine article on what was then a worldwide concern over "global cooling":

TABLE A3-1. CLIMATE CHANGE IN THE

Time Frame	Perceived Trend	Sample	Media Sources
1895–1920s	Global Cooling	"Ice Age"	NY Times, LA Times, Chicago Tribune
1930s	Global Warming	"Dust Bowl"/Hottest Year on Record, 1934	Many sources.
1950s	Global Cooling	"Climate: The Heat May Be Off" (1954)	Fortune
1960s	Global Warming	"Arctic Ice Will be Gone in Two Years" (1969)	NY Times
1974–1980s	Global Cooling		Fortune, Time, Newsweek, Science News
1988–2010	Global Warming		Many sources

> Record rains in parts of the United States, Pakistan and Japan caused some of the worst flooding in centuries. In Canada's wheat belt, a particularly chilly and rainy spring has delayed planting and may well bring a disappointingly small harvest . . .

The atmosphere has been getting gradually cooler for the past three decades. The trend shows no signs of reversing. Climatological Cassandras are becoming increasingly apprehensive, for the weather aberrations they are studying may be the harbinger of another ice age.

Man, too, may be somewhat responsible for the cooling trend. Climatologists suggest that dust and other particles released into the atmosphere as result of farming and fuel burning may be blocking more and more sunlight from reaching and heating the surface of the earth.

The record of past climate change alarmism is easily applied to modern times. Today we see the same attempt to blame mankind for global warming, just as was done during the cooling period of the 1970s with the same pollution-from-industry story line.

Similarly, as this cold climate extends its control over the Earth, we should not allow yet another bout of climate hysteria to cause people to think that we are heading for a new ice age. Some respected scientists have suggested this. However, I do not believe there is sufficient evidence to indicate, from a solar cycle standpoint, that we are heading into a climate period worse than a Dalton Minimum class of cold. Given the tenuous nature of the world's global food supplies and the percentage of Americans and other citizens around the world who are already underfed, the forecasted cold for the next two or three decades will be bad enough. If these other scientists are correct and we enter a Maunder-class cold period, then the human suffering resulting from the loss of crops will be biblical in its scope.

Another recurring error made by both sides of the man-made global warming issue is the overemphasis of very short-term temperature spikes or declines of just a few years or less, when it is the long-term trend that determines a change in climate. Small-duration, yet attention-grabbing events like the global temperature decline of 2007–2008, the record-setting snows and cold that struck the US capital during the winter of 2009–2010, and the heat of the El Niños of 1997–1999 and 2009–2010 cannot stand by themselves as sole determinants of a change in the direction of Earth's climate. They can have meaning toward a new climate change only when viewed in the context of a longer-term global temperature trend and its associated relational solar activity cycle.

REASON 8:

Cold temperature records are being set, globally.

There are many real signs that the long-term heating of the Earth has ended just as the Relational Cycle theory says it should, and that land temperatures, ocean temperatures, and hurricane patterns are returning to normal, if not distinctly colder climate patterns. To be sure, heat and cold records are set every year in some corner of the world. These "anomalies" are always followed by meteorologists and climate scientists and don't necessarily need scrutiny unless we start to see larger numbers of significantly warmer or colder records being set. During the 1990s and 2000s, we saw the warmer side of these anomalous readings. Now that the hibernation of the Sun has begun, we have seen a dramatic reversal in such temperature records that are fully consistent with the predicted change to a colder climate. Recent signs that change is under way began in 2007, including the following:[4]

- From December 2006 through January 2007, Malaysia had its worst flooding in decades.
- In January, winter storm Kyril hit the British Isles and Europe, with 170 KMH winds and 50 lives lost.
- In January, Bangladesh had a cold wave — the coldest in 40 years.
- In January, Brazil had heavy rain and flooding.
- In January and February, Bolivia had heavy rain and flooding, affecting 200,000 people.
- In February, a major winter storm affected over 300,000 in the northeastern United States and southeastern Canada.
- In March, China had its heaviest snowfall in 56 years.
- In May, Uruguay had its worst flooding since 1959, with more than 110,000 people affected.
- In May, Argentina reported below freezing temperatures, resulting in gas and electric shortages. In June, Buenos Aires had its first major snowfall since 1918.

- In May, South Africa reported 54 cold weather records, and in June, Johannesburg received its first significant snowfall since 1981.
- In May and July, record-setting rains and flooding hit China.
- In August, Zurich, Switzerland, had its largest daily rainfall in a hundred years.
- In November, Switzerland had its heaviest snowfall in 52 years.

More Recent Years of Cold Weather Events

Now let's fast-forward to more recent years to see whether these strange cold weather events have continued. The answer is a resounding yes! During the early winter of 2009–2010, many cold weather records were broken in the Northern Hemisphere. Here are but a few examples:

- **In the first week of December 2009, Houston had its earliest snowfall on record.**[5] This snowfall broke the standing record, which was set only the previous December. From January 8–10, Houston had its coldest temperatures in 14 years.[6]
- **In 2009, Europe was in the grip of one of the coldest winters in decades, including 12 inches of snow in the UK.**[7]
- **China: Beijing had its coldest (near 0°) day in 50 years, along with its worst snowfall in decades.**[8]
- It was reported on Houston radio (WFVB) that on January 8, 2010, Bismarck, North Dakota, had a temperature of -38°F and a wind chill of -50°F. These are temperatures normally found in Antarctica and the North Pole!
- **On January 11, 2010, KUHF and NPR reported that Miami, Florida, had a record cold temperature of 37°F, breaking an 83-year-old record from 1927.**

- **Historic record snows of February 2010 paralyzed the nation's capital.** Who can forget the widely reported and amazing snowfalls that struck major eastern cities, setting all-time record depths of snow in Washington, DC, Baltimore, and Philadelphia?

And what about the winter of 2010–2011?

- On November 14, 2010, Minnesota had its largest snowfall in decades, according to The Weather Channel. **Winter didn't officially start for at least another month, on December 21.**

- On December 2–3, 2010, an early snowstorm strangled parts of New York, Ohio, and the Great Lakes region, according to The Weather Channel, with some people stranded up to 20 hours on I-90. **The Weather Channel reported that as of the week of February 6, 2011, there had already been eight major winter storms.**

- The community of International Falls, Minnesota, which claims to be the coldest city in the United States, outdid themselves. On January 21, 2011, they set a new record cold of -46°F, beating the old record handsomely. It was previously -41°F (1954), as stated on The Weather Channel.

- **As a general comment, the winter of 2010–2011 was one of the coldest and certainly one of the snowiest on record in the United States.** A simple web search will show hundreds of references, if anyone really needs them. Most Americans from Oklahoma to Boston are well aware and will say, "Don't bother." To save you the time, here are just a few you will find: My hometown of Orlando had its coldest December ever (Channel 13 News). Hartford, Connecticut, had its snowiest month on record for January. As of February 3, 2011, Atlanta had four times its average snowfall. Dallas had its coldest weather in 15 years, in the first week in February. Tulsa, Oklahoma, had its greatest one-day snowfall ever, and Oklahoma City had more snow in 11 hours than it sees in an average year, both in the first week of February. By that same week, New York City had five times its average yearly snowfall, and that was before midwinter had arrived! Credit for these last few items goes to Paul Douglas at his Saint Cloud, Minnesota, Blogspot.

- A powerful cold front struck Europe in the first week in December, causing snow and bitter cold from Ireland deep into Eastern Europe. On December 2–3, temperatures reported by Fox News went to 5°F in Poland, where 30 people died from its effects.
- **During the same early December period, Sweden reported its coldest weather in a hundred years!**[9]
- **Polish weather experts predicted Europe and Russia will have to endure its coldest winter in 1,000 years!**[10] While the early harsh winter that descended on Europe seemed to back this ominous prediction, other sources disagreed with the forecast, including me. Still, it gives pause to reflect on the many years of opposite and false predictions we have been subjected to that our planet, and therefore its seasons, can only get warmer and warmer. The cause for this extreme cold weather prediction was apparently a declining flow and lower temperature of the Gulf Stream that warms Europe. This is a truly serious situation for Europe and Russia and the rest of the planet should it unfold in the future. Here in March of 2011, the forecast seems to have missed the mark by a wide margin. What if these meteorologists missed it by just one year and such a climate cataclysm hits Russia next winter or the one after?

- **In what some might call a case of divine intervention, at the site of the December 2010 UN COP 16 climate conference in Cancun, Mexico, cold temperature records were set six days in a row.**[11] This temperature report was one of many where new cold records were being set and, of course, rarely made it to the newspapers or TV news. Unfortunately for Earth's inhabitants, the Cancun climate conference had some modest success toward arriving at a framework to replace the Kyoto Protocol, mostly because of limited goals at the outset.[12] I say "unfortunately" because once more the United Nations Intergovernmental Panel on Climate Change and global warming supporters in other countries have managed to keep alive the myth of man's ability to change Earth's climate unilaterally, diverting attention away from the coming cold weather onslaught and in turn ensuring more people than necessary will have to suffer the consequences of global turmoil it will likely produce. In the midst of a concurrent record deep freeze that at the time was striking the United States and Europe, another global warming zealot, former US vice president Al Gore, was undeterred by reality and still beating the AGW drum in stark contrast to the record cold. He said, "I'm a little depressed about Cancun. The problem is not going away. It's getting steadily worse."[13] Doubtless the problem he was referring to was his lack of cold weather clothing on his highly paid globetrotting tour to push what more and more brave scientists are saying is the greatest fraud of modern science.

Why all the statistics about rains and flooding along with cold temperatures and snowfall? What is generally found, according to my studies, is that there is a general increase in rainfall and major flooding events during major cold climate eras. For corroborating evidence of this, check NOAA precipitation records versus cold temperature records and then compare them to sunspot count.

- **Britain recorded its coldest December in 120 years.**[14] This astounding record almost became the worst in 300 years.[15]

- **South Korea recorded its largest snowfall in a hundred years.** On February 14, 2011, The Weather Channel reported on South Korea's record snowfall.

This winter of 2010–2011 was unusual and clearly a global event. It is an appalling insult to all of the world's citizens that we continue to see news reports and hear so-called "experts" blame man-made global *warming* for record snowfalls coupled with record cold temperatures. It is but another indicator of how AGW is the greatest scientific fraud in the history of modern science and displays repeatedly the unrestrained arrogance of its authors and supporters.

There are tangible signs that Earth's oceans have also started to cool again. The Sun-caused heating of the oceans, which has been going on for the past few decades (and longer), has stopped. The past trend of ocean warming is reversing to a cooler phase (or has already reversed in some oceans). Here are a few pieces of evidence:

REASON 9:
A NASA ocean temperature study has shown declining ocean temperatures.

A study by Dr. John Lyman of NASA,[16] with a later correction by another team member, has demonstrated that the oceans are cooling. Lyman's initial paper showed meaningful reductions in the temperature of the world's oceans. He received criticism for such AGW heresy, and his collaborator, Dr. Josh Willis, subsequently released a correction[17] based upon the fact that new ocean buoys, used for data gathering and deployed globally, apparently had biased temperature equipment. However, I find Lyman's general conclusion is still valid. My review of the published correction still showed that the heating of the world's oceans had in fact *stopped* in any case. A check on current ocean temperatures shows Lyman was right all along.

REASON 10:

A review of ocean temperatures measured by satellites shows a cooling trend.

A cursory review of the extent of cooling versus heating for a recent 10-year period shows a clear cooling trend in ocean temperatures. This became quite obvious with the posting of the ocean temperature chart at the Dr. Roy Spencer site (www.drroyspencer.com), which included data up to December 2010. The chart was striking in how it showed the steepest drop in ocean temperatures in over ten years.

The ongoing record drop in the ocean temperatures is fully in line with my latest prediction of a historic drop in the world's temperature projected to arrive before November–December 2012 and that it will be greater than that seen between 2007 and 2008.

Since this prediction was first made, there has been ample time to evaluate whether the predicted average global temperature drop of 2012 exceeded that of 2007-2008. A March 2014 review of global temperature charts from the leading institutes that track this parameter shows the following:

	Institute	**Met Prediction?**
1.	University of Alabama Huntsville	No
2.	Remote Sensing System	Close match
3.	Goddard Institute for Space Studies (GISS)	Yes
4.	Hadley Center/U. of East Anglia (HADCRUT4)	Yes

The chart below from the from one of these institutions, the GISS, shows that the average global temperature line in 2012

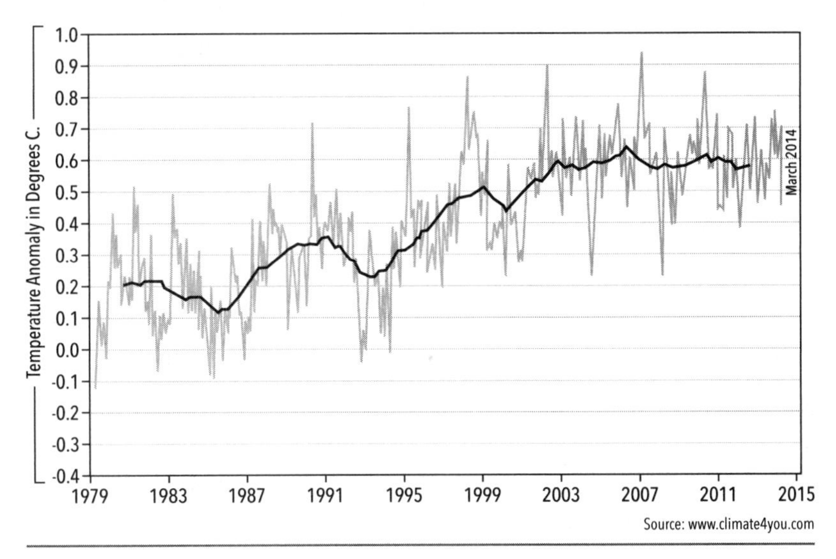

Figure A3-4. This chart from the from one of these institutions, the GISS, shows that the average global temperature line in 2012 dipped below that of the 2007-2008 period confirming the predicted drop.

dipped below that of the 2007-2008 period confirming the predicted drop.

By the way, continue to check the reliable sources found in this book for global temperature data like that of Dr. Humlum and Dr. Spencer. When the next temperature record is set, it will be unlikely, as before, that you will read about it in the papers or see it on the evening news! If you do learn about it, the odds are this much colder temperature will somehow still be blamed on man-made global warming!

REASON 11:

Ocean cooling is predicted by a California seal pup study.

A multidecade study of seal pup weights has shown that they increase during colder periods. Seal pups have begun to reach heavy weights again. From the report by Burney Le Boeuf and D. E. Crocker:

> By way of Dennis Avery, "After 1999, however, the ocean temperatures began to decline, fish became more abundant, and the pups' weaning weights abruptly began to rise. By 2004 the pups' weaning weights had recovered to 90 percent of their 1975 weaning size."

I include this particular quote not as anecdotal evidence of the cooling oceans but rather to say if the natural world, the plants and animals, are telling us it is getting colder, then it probably is.[18] Doubtless we could find many other such studies with a little effort where Earth's flora and fauna are sending cold climate signals.

REASON 12:
The Pacific Decadal Oscillation (PDO) has started.

One of the most telling measures of the state of the ocean and whether we are in for a long climate change is found in the Pacific Ocean. Here there are two major temperature cycles that scientists closely follow to determine future regional and global climate and meteorological events.

The first is the El Niño Southern Oscillation (ENSO). This phenomenon of temperature variation of the equatorial waters west of South America has two components: the warm El Niño phase and the cool La Niña phase. These cycles of Pacific Ocean temperatures can last for 6 to 18 months and occur every 3 to 20 years, directly affecting the climate of the United States and other countries.

Another powerful system, the Pacific Decadal Oscillation, brings cold waters to the west coast of the United States on a much longer cycle. Dr. Don Easterbrook, Department of Geology, Western Washington University in Bellingham, has come to an important conclusion. Based upon his own research and NASA satellite data, he stated in July 2008 that the PDO will be in place for the next 25 to 30 years, which means that "the global warming of the past 30 years is over." The last time this occurred, according to Dr. Easterbrook, was during the 1940s, and it caused 30 years of global cooling.[19] Figure A3-5 explains the PDO further.

Dr. Easterbrook's announcement followed quickly on the heels of a NASA announcement of a major change in the PDO in April 2008 and my own "end of global warming" statement on July 1, 2008. The NASA announcement clearly indicated that a shift to a long cool PDO phase had begun:[20]

"The shift in the Pacific Decadal Oscillation, with its widespread Pacific Ocean temperature changes, will have significant implications for global climate. It can affect Pacific and Atlantic hurricane activity, droughts and flooding around the Pacific basin, marine ecosystems and global land temperature patterns."

The PDO shift to a long cool phase has gained wide acceptance. This may also be a result of NOAA finally biting the bullet and admitting the Pacific may be much colder for many years to come. Figure A3-5 shows the variations between warm and cold

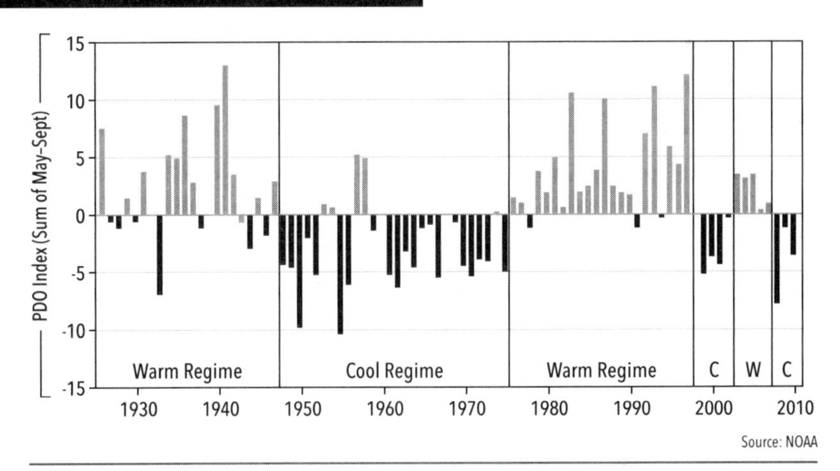

Figure A3-5. Pacific Decadal Oscillation (PDO). Time series of shifts in sign of the PDO, 1925 to 2010. Values are averaged over the months of May through September. Bars above the "0" line indicate positive (warm) years; bars below the "0" line indicate negative (cool) years. Note the beginning of the predicted new long cold phase to the far right of the chart, which began 2009-2010.

phases of the PDO from 1925 to 2010. The most recent ten years in that span saw rapid changes between cold and warm phases. The last change to a cold phase, shown in the bars at the far right of the chart, which began in 2009–2010, is the one we should be concerned about. It is the one Dr. Easterbrook, NASA, and others have focused on that allows a forecast of decades of a new colder climate.

Even one NOAA source now says as much by the assessment, "It is thought that the PDO has dipped into the negative (cold) phase and may remain in this phase for another 10–20 years."[21]

The last time this occurred, as seen in Figure A3-4, was the period from the mid-1940s to the mid-1970s, when the Earth experienced 30 years of colder weather.

There can be little doubt that the warming of the Earth's oceans has at least stopped. From these few yet important pieces of evidence, it appears they have started cooling, and may be much colder for decades into the future. As the oceans cool, so will the rest of the planet in turn. Again, for those willing to examine the facts and set aside the conventional thinking and political correctness of our day, there is only one reasonable conclusion we can make about the Earth's climate: global warming has ended and a new cold era has begun.

The World's Glacial Ice Is Growing

The majority of the Earth's glacial ice fields are once again growing and apparently have been for ten years or more in most areas.

Again, this statement begs the question: Why haven't we been told this? Or, better yet: Why are we still being told that the world's glacial ice is melting? Why has the US government recently put out a report for Congress saying that the world's glaciers are melting if in fact they aren't? In summary, here is what we know about the world's glacial ice in the three areas where it exists: Antarctica, Greenland, and in the mountains.

The Antarctic Glacial and Sea Ice

REASON 13:

Antarctica is getting colder, and its glacial ice sheet is growing and will continue to do so for the foreseeable future.

Several important studies have been done that back up this claim, including those by Peter Doran,[22] Curt Davis et al.,[23] IPCC,[24] and Marco Tedesco and Andrew Monaghan.[25]

Here is a quote from the Doran study, published in 2002. The study, done for the National Science Foundation (NSF), shows that Antarctica has in fact been on a long-term cooling trend for many years:

> "Our 14-year continuous weather station record . . . reveals that seasonally averaged surface air temperature has decreased by 0.7°C per decade. The temperature decrease is most pronounced in summer and autumn. Continental cooling, especially the seasonality cooling, poses challenges to models of climate and ecosystem change."

After being harangued by the AGW forces (I believe), Doran attempted to clarify his team's conclusions in a *New York Times* op-ed three years later. I read it. His attempt to placate those accusing him of being against AGW was well written. He did make clear that his study was being taken too far by AGW skeptics, but he still stood by the methodology. The findings are what they are, and their conclusions still stand. I have never met him, but my take on Dr. Doran is he is one of many outstanding researchers out there doing great science but whose work will be exploited by both sides in the climate change debate. Unfortunately, all science on climate change these days is being categorized as politically correct or not, instead of just being sound, professional research or not.

From the study by Curt Davis and others, when referring to their satellite survey of Antarctica from 1992–2003, they found that Antarctica had gained 45 billion tons. His comment: "It is the only large terrestrial ice body that is gaining mass rather than losing it."

IPCC, in its last report, AR4 (2007), finally accepted the truth by conceding the Antarctic continent will be gaining ice for the foreseeable future:

> "Current global model studies project that the Antarctic ice sheet will remain too cold for widespread surface melting and is expected to gain in mass due to increased snowfall."

I am sure that few if any of you have ever seen this shocking admission from IPCC, much less read it on the front page of a newspaper or heard it on the evening news. Not to be totally conciliatory, the UN experts added a defensive and almost laughable caveat so AGW supporters would not scream too loudly:

"However, net loss of ice mass could occur if dynamical ice discharge dominates the ice sheet mass balance."

Our beloved Yogi Berra could not have said it better. Or did they have him write this?

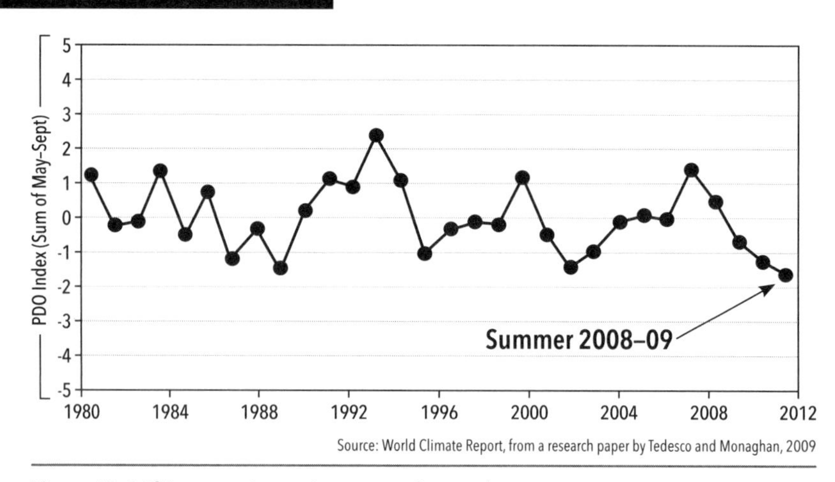

Figure A3-6. Thirty-year Antarctic snow melt record.

I interpret the UN's hedge as, "We at the UN really hope we are wrong and that Antarctica does start to melt again soon so we can have the flooded coastlines we have been predicting for 20 years. How else can we scare the people into doing what we want?"

The final piece offered in this short assessment of what is happening with the Antarctic ice sheet is provided in an important study by Marco Tedesco and Andrew Monaghan from 2009. Tedesco, by the way, has a record of reporting on ice melting and not growing. Figure A3-6, from his study, shows unequivocally that Antarctica is not melting.

This remarkable chart has much relevant information:

- It shows that Antarctica had recently established a 30-year record for the lowest amount of summer snow/ice melt! In other words, Antarctica has been getting much colder and has been doing so for a long time.
- This is part of a long-term trend that can be seen from the chart, which began at least by 1992 and possibly earlier.
- It shows again how the media and US officials continue to quote only those climate events regarding Arctic ice melt that support their flawed AGW policies and ignore or cover up even more significant events (i.e., Antarctic ice growth) that don't support their deception. Lest we forget, Arctic sea ice melt has no impact at all on sea level rise, whereas Antarctic glacial ice has everything to do with it.

World Climate Report Chief Editor Patrick J. Michaels and team have done an outstanding job bringing out this stunning news and have their own assessment of why this supports my belief that we can't trust our government to tell us the truth about climate change. Here is Patrick's assessment of the Tedesco report from his October 6, 2009, posting at www.worldclimatereport.com:

"The silence surrounding this publication was deafening.

> "But this time around, nothing, nada, zippo from NASA when their ice melt go-to guy Marco Tedesco reports that Antarctica has set a record for the lack of surface ice melt (even more interestingly coming on the heels of a near-record low ice melt last summer).
>
> "So, seriously, NASA what gives? If ice melt is an important enough topic to warrant annual updates of the goings-on across Greenland, is it not important enough to elucidate the history and recent behavior across Antarctica?"
>
> Good for you, *World Climate Report!*

Here once more we see the grand deception being played on us. There is unequivocal science showing Antarctica, a massive continent that contains 90 percent of the world's glacial ice, is setting records for cold weather and glacial ice growth. Yet this vital information that should be a fountain of relevant knowledge in the climate change debate is being intentionally kept from the American people.

Here again, Patrick Michaels' *World Climate Report* steps forward to clear the air and tell us the truth. How distressing this situation is. To get the truth about our climate, we need to turn off the TV, throw our newspapers in the trash, don't believe a thing our own government is telling us about the climate, and instead go online!

Where would we be without the Internet? Yes, I know. It has viruses to contend with, annoying pop-up ads, people stealing information out of your computer for their marketing programs, and just as many rumor and lies as it has truth, but at least I know "the truth is out there." The Internet is far from perfect. It is both light and darkness, but at least there is some light. Regrettably, among some of our government science agencies, the White House, and some media sources, that light has gone out.

The Greenland Ice Sheet Status

REASON 14:

The Greenland glacial ice sheet is stable and growing.

Just like Antarctica, in order to understand Greenland, we should start by looking at the temperature profile. Is there any evidence that Greenland is getting colder? Again, the answer is yes! Are we being told about such proof? Again, the answer is no!

Edward Hanna and John Cappelen reported in 2003 that Greenland, especially southern Greenland, was cooling.[26] Here is an excerpt from their findings:

> "The Greenland air temperature data showed a cooling of 1.29°C over the period of study (1958–2001), while two of the three [sea surface temperature] databases depicted a cooling of 0.44°C and one of them cooling of 0.80°C."

It was further reported in their study that "recent cooling may have significantly added to the mass balance of at least the southern half of the [Greenland] ice sheet."

I have read many research reports in the course of writing this book. One of the very best is that of B. M. Vinther et al., published in June 2006.[27] This impressive piece of solid, objective science has some unique yet foundational findings with regard to the long-term temperature record for Greenland from 1784 to 2005. Using a variety of sources, which they merged into a total record and in-depth mathematical analysis of raw data, they came to the following revealing conclusions:

> "The warmest year in the merged record is 1941, while the 1930s and 1940s are the warmest decades . . . The coldest year is 1863, while two cold spells (1811 and 1817–1818) make the 1810s the coldest decade . . . The marked cool periods in 1811 and 1817 follow two large volcanic eruptions in 1809 (unidentified) and 1815 (Tambora)."

The significance of these and other conclusions in the report verify the following:

> Using data to 2001, the warmest period in Greenland occurred in the pre-World War II era, before the massive global industrialization of the last 70 years and before the emission of most greenhouse gasses, the alleged cause of anthropogenic global warming.

The concept of man-made global warming is invalidated by the evidence from historical temperature records in Greenland!

The center of the Dalton Minimum, the last solar hibernation, was the decade of the 1810s, and the cold bottom of the last Bicentennial Cycle was the coldest in Greenland since 1784. It was made even colder by two volcanic eruptions.

These conclusions and others within the report support my own opinions:

- Greenland has been subject to frequent and significant temperature fluctuations. Therefore, isolated glacial rebuilding or melting events can vary widely within just a few years. This means that short-term variations are normal and cannot be extrapolated or reliably point to either a cooling or warming trend. An excellent example is the recent evidence of cooling demonstrated in the Hanna, Vinther, and related studies, yet there are other studies of the years during the same decade that show rapid ice melt along the coasts, implying a warmer Greenland. My objection to the reporting on Greenland ice studies is that the general public and our leaders rarely hear of Greenland being colder or that its ice sheet is growing. That side of the climate is missing, unreported. A balanced reporting of what is going on in Greenland is just not happening. On the other hand, any study that shows the slightest hint of melting ice, even if it is normal summer melt and glacial calving, gets maximum coverage.

The endless propaganda by AGW extremists has convinced a generation that seeing TV news reports or documentaries depicting normal, seasonal glacial shedding of massive sheets of ice into the sea is a clear sign of man-made global warming at work.

- Greenland weather and temperature events are not directly comparable to Antarctic or mountain glacial events and, in fact, can be quite different because of its size, its position in the North Atlantic, its span of Northern Hemisphere latitudes, its proximity to the Arctic Circle, its exposure to the warm waters of the Gulf Stream, and the influence of the North Atlantic Oscillation (NAO), predominant wind patterns, and other related North Atlantic weather phenomena. Greenland effectively has "one foot in the freezer" on its northern Arctic coast and "one foot in the hot tub" of the Gulf Stream at its southern coast. For comparison, the southern tip of Greenland is roughly on the same latitude as Oslo, Norway; Stockholm, Sweden; Helsinki, Finland; Leningrad, Russia; and Anchorage, Alaska.
- Greenland's temperature over a recent 20-year period, including the record global warmth of 1995–2005, is well within the normal range of variance compared to the mean over the last 150 years! In other words, Greenland is doing just fine, thank you.
- Whatever the world is doing temperature-wise, according to the data, Greenland seems to march to its own drummer — its own set of regional influencing factors — and is otherwise effectively independent of any external effects of man.
- It appears, however, that Greenland does react to the Relational Cycles of the Sun, especially the Bicentennial Cycle and the effects of random volcanic activity, once again

lending specific credence to the use of the Relational Cycle theory as a climate monitoring and forecasting tool.

So that is the big picture for Greenland's temperatures. Now, what about the question of whether the Greenland ice sheet is growing? What you haven't heard is that the interior of Greenland's ice has been on a growth trend, and its recent growth trend may have started as far back as 1992!

Have you ever seen that on the front page of your local paper or on the evening news? Chances are that has not happened anywhere in the United States. Who is it that has systematically kept this information from us, and why are they doing so? Why is it that we have been intentionally misled about the status of Greenland's ice shelf for 15-plus years? These are sad times for science and equally sad times for our citizens, when they are given only one side of the important climate change story — the only story taught in our schools — and the rest of the science is purposely hidden, obscured, covered up.

Where is the evidence that this assertion, that Greenland's ice is growing, is fact? Covering the period 1992 to 2002, Dr. H. Jay Zwally et al. reported in a 2005 article in the journal *Glaciology* that the Greenland ice sheet had at that point been growing by a net 11 billion tons per year, plus or minus three billion tons per year.[28] Dr. Zwally now says the situation has reversed, but how is it that Greenland's ice could have gained mass while global temperatures were increasing and, in 1998, reached their highest on record? Greenland, it seems, has its own mind about what it will and will not do.

Are there other studies we have that can give us more perspective on Greenland?

Yes. Publishing in the journal *Science* in 2005, Dr. O. M. Johannessen et al. reported that snow accumulation had increased in the interior of Greenland between 1992 and 2003.[29] The amount of increase was 6.4 centimeters each year above 1,500 meters elevation. Below that elevation, it was reducing by only 2 centimeters per year.

These are but two of many studies performed by experts from many countries. The picture with Greenland is complex. The edges of Greenland can be melting while the interior is growing. Glacial flow processes are not yet fully understood. Some of the best science on Greenland has only recently started, using better satellite data. Even the best scientists studying this island say they need more observation and data over longer periods. Greenland's location makes it subject to a host of factors that mountain glaciers, the Arctic, and Antarctica don't have to contend with. What they all share, however, is that what we have heard and read about them for the last 20 years is quite different from the reality.

According to researchers, at least to 2005 Greenland had undergone a net significant increase in its ice sheet for at least 16 to 18 years! Why have we been led to believe that Greenland has been melting since modern global warming began? Why have we not heard of the important revelations about the temperature record of Greenland? I believe the rapidly approaching cold climate will dramatically change the picture for Greenland, along with the rest of the world.

The Arctic Sea Ice Status

There is little doubt that the Arctic sea ice has been in decline for almost two decades. That is entirely expected, given the peak heating of the 206-year cycle caused by the Sun. With the advancing cold era, that will change, and initial signs of a coming change are already visible.

REASON 15:

During periods of time in 2009 and 2010, the Arctic sea ice reached its historical average extent.

The typical story the public has been fed over the years is that the Arctic is melting rapidly and that soon it will have no ice whatsoever. A relatively new group, initially organized by nine European countries, is the Arctic Regional Ocean Observing System (Arctic ROOS); its associated research institution is the Nansen Environmental and Remote Sensing Center (NERSC) in Bergen, Norway. Their regular tracking of Arctic ice provides an objective resource for anyone who no longer trusts what they are being told about the status of Arctic sea ice. Their research has shown that for 2009 and 2010, the Arctic sea ice area was near or within its 1979–2006 average range for a period of months.[30] Why weren't we told this? Why was this good news not on the front page of leading newspapers or the lead item on the evening news, or better yet, why didn't we see the NOAA administrator call a news conference to announce that we no longer need to worry about a melting Arctic and that things may be returning to normal? The polar bears have been saved!

With the coming solar hibernation, the past Sun-caused trend of reduced Arctic sea ice is expected to reverse significantly, with the Arctic setting all-time records for sea ice extent in the next two decades. Yet we continue to see only a one-sided view of the Arctic from supposed government experts and celebrities. For example, here are some quotes from the National Snow and Ice Data Center (NSIDC) and Al Gore on the Arctic sea ice:[31]

- In April 2008, Dr. Mark Serreze from NOAA's NSIDC said there would be an ice-free Arctic that same year.
- When that didn't happen, he gave another prediction that "the Arctic would be free of summer ice by 2030."
- In 2008, Al Gore said that the "entire North Polar ice cap will be gone in five years." Didn't he also invent the Internet?

The 2009–2010 Arctic ice extent records don't seem to be helping Gore's cause. From one report by David Rose (from www.dailymail.co.uk on January 10, 2010), we have the following statement that I can just about guarantee no one saw on the front page of any newspaper in the United States: "According to the US National Snow and Ice Data Center in Colorado, Arctic summer ice has increased by 409,000 square miles, or 26 percent, since 2007 — and even the most committed global warming activists do not dispute this."

What was that? Arctic sea ice has been growing rapidly for three years and we weren't told? How is that possible in the land of free speech? And why, with this evidence, do we still see stories about the Arctic sea ice melting away to nothing? We should all be appalled that outlandish, irresponsible statements like the above forecasts for an ice-free Arctic made by prominent individuals can be given the slightest credence, print space, or air time. Yet because such predictions fit within the political framework of the day, they are acceptable and proper.

There has been much speculation over the past decade about whether the melting of Arctic sea ice would open up new sea lanes for shipping and possibly permit drilling for oil and gas reserves. All these plans are for naught, and those nations and companies that are sinking substantial amounts of capital into these "ice-free Arctic" ventures will likely see only negative results. The new cold climate will put a "freeze" on such plans and will do so quickly! I predict the Arctic will set all-time records for sea ice extent in the next two decades, and the change to this period of growing ice will become obvious within a few years.

I am hardly the only one with such a strong opinion on the future for Arctic sea ice. There are others, especially outside the United States, where scientists are free to speak their mind without fear of retribution from AGW supporters and the politically correct establishment that rules the roost here in the United States. Take this statement, for example, from Dr. Oleg Pokrovsky at the Russian Voeikov Main Geophysical Observatory. As reported April 23, 2010, by www.ferratermora.org, he said, "Politicians who placed their bets on global warming may lose the pot." In referring to a return to the Arctic cold of the 1950s and 1960s, he said the coming cold will peak in 15 years, and "the northern passage will freeze and it will be impossible to pass through it without icebreakers."

Polar Bear Politics

Since we are on the subject of the melting of Arctic sea ice and its effects, this is probably a good point to bring up the matter of the polar bears or, rather, what I call "polar bear politics."

One of the greatest lies knowingly forced upon us by AGW deceivers during the past period of global warming is that polar bears are on the verge of extinction. We have been bombarded by photos and videos of lone bears swimming among small Arctic ice floes or empty seas, implying that melting Arctic sea ice is causing them to drown or starve for lack of ice to hunt and rest on. Yet not one drowning or death has been proven to be the cause of greenhouse gas emissions. These powerful animals are capable of swimming many miles in search of prey; they are thriving. Yes, you read it correctly: they are thriving! Even if all of a sudden we could confirm that polar bears are dying from hunger, with the size of their present estimated population, statistically we would have to start seeing hundreds of dead bears, with confirmed causes of their demise, before it would constitute a statistically meaningful trend. And that simply isn't happening.

When it comes to polar bears being threatened, we have been scammed!

Many organizations use these distressing images of polar bears and claim they are under threat of extinction in order to push "green" products for sale or, in the case of environmental groups, pry donations from people. Our schoolchildren have been falsely led to believe that saving the polar bears is a necessary and urgent goal. We have been force-fed stories of polar bear travail as a key reason for supporting all green programs or even buying an "environmentally friendly" car. But what is the real truth about their condition?

The truth is that since the mid-1960s, when their population was estimated at about 10,000, polar bears now number well over 20,000![32] Further, their numbers since 1972 have been in a relatively stable range of between 20,000 and 25,000, and that includes during the peak decades of the Earth's Sun-caused global warming between the mid-1980s and the present.

An important but undisclosed issue surrounding the entire game being played upon us by AGW enthusiasts is we just don't know how many polar bears there are!

In order to find out the current status on polar bears, I went to the Polar Bear Specialist Group (PBSG). They claim to be "the authoritative source for information on the world's polar bears." This group is comprised of members from all those countries with polar bears, including the United States, and is a subordinate organization of the International Union for Conservation of Nature (IUCN). The IUCN, according to their website at www.iucn.org, is "the world's oldest and largest global environmental network — a democratic membership union with more than 1,000 government and NGO organizations, and almost 11,000 volunteer scientists in more than 160 countries." US members include organization and agencies ranging from the World Wildlife Fund to the US Fish and Wildlife Service.

In their most recent 2013 survey, the PBSG said that at least 18,349 polar bears exist based on surveys in 19 Arctic zones under consideration.[33] Of those 19 zones, the PBSG says the populations are seeing a reduction in only four zones, are stable or increasing in six zones, or in most cases (nine zones), they just don't have enough data to know one way or the other. This is an astounding admission. The PBSG says in effect that in 49 percent of the areas where it actively monitors polar bear populations, it simply doesn't know how many there are, much less whether they are increasing or declining.

In all cases, just like the IPCC reports, the effects of melting Arctic sea ice that are postulated to cause a serious threat to the polar bears in the distant future are based on the flawed AGW greenhouse gas theory and apparently incomplete data. They are certainly not based on the repeating warming and cooling cycles of climate change by which the Sun operates.

So if polar bears aren't threatened by declining Arctic sea ice, and their numbers have been healthy (if not growing) during the last decades of global warming, then what *is* the threat to polar bears?

I believe there are now five real threats to polar bears:

1. Their numbers have grown during the last decades of global warming, to the point that they may be at risk because of overpopulation (i.e., global warming has been too good for the polar bears).

2. They are in fact being threatened and harassed as a result of the many well-funded global warming and conservation studies and the hundreds of researchers, media types, and nature photographers that constantly harass them by flying over them with helicopters and planes, or invade their habitat with noisy trucks and boats and then shoot them with drug-filled darts, tag them, and take tissue and blood samples for analysis. I am not a biologist, but I have to think these bears and their scared-to-death little cubs at their side aren't comforted by the humans with guns saying, "They'll be just fine when the drug wears off."

3. The misguided opportunity that several nations have seen for oil exploration in the Arctic, with its falsely predicted disappearing sea ice, has caused large ships to begin to break through the Arctic ice in ever-increasing numbers. How do you think bears and their cubs (or, for that matter, the seals the bears live off of) will fare when these huge, noisy steel monsters begin to routinely plow through their

neighborhood, breaking the ice with ferocity as they create and then try to keep open a new Arctic passage?

4. The combined effect of (2) and (3) above may be causing real risks to the normal behavior and mental stability of polar bears, socially, healthwise, and perhaps reproductively. Wild animals under constant harassment and stress from humans don't do well, right? I've got it! How about a polar bear study to study the effects on polar bears of polar bear studies.

5. Their numbers will again decline because of the next cold climate era. Polar bears, according to their census records for over 50 years, have lower numbers during extreme cold periods and explode as a population during warm periods. As a result, they may not do well in this new solar hibernation.

Wait a minute. I have an even better idea. Perhaps what we should do is form a new Polar Bear Protection Society (PBPS). Its charter will be to protect polar bears from government, environmental, and conservation organizations whose mission is to protect them. What do you think? If we see any researchers harassing (I'm sorry, I meant to say "protecting") the bears, we will shoot the researchers with drug-filled darts, take tissue and blood samples, and, oh yes, confiscate their expensive camera gear and pilot's/boater's licenses. For any researchers we catch that are funded by US taxpayers, we will put heavy GPS tracking collars on their necks so we can follow them on their nefarious meanderings within the Arctic Circle. When the battery starts to run down and we get a low-voltage warning signal, we can always track them down quickly and tap them with another dart. No big deal. Besides, they'll be just fine when the drug wears off.

Yes, I'm being facetious, but you get my drift.

What I call "polar bear politics," which we have had to suffer through for two decades, is but one of many sins of the political correctness of the past climate era.

Mountain Glacial Ice Status

Similar to the Arctic sea ice story, mountain glacial ice has also seen many years of reduction in the "glacial ice mass balance" in keeping with the Sun-caused heating from the 206-year cycle. This has also now reversed.

REASON 16:

The world's mountain glacial ice began to reverse from a predominantly shrinking phase to one of long-term growth beginning around 1998.

I have examined widely quoted sources for mountain glacial ice status and researched numerous papers on the subject. A review of the glacial ice records from the Department of Geology, University of Zurich, Switzerland, for example, shows growing mountain glacial ice. Unfortunately, I also found the Swiss database wholly inadequate to make a proper assessment of the world's glacial ice status! It should not be used by governments or anyone else to determine global policy or the status of glacial ice. The majority of glaciers for which it maintains data are located in Europe; therefore, it should only be consulted for European glacier data. Nevertheless, according to their publicly available data, while there has been an overall decline of glacial ice for the past hundred years, beginning around 1998–1999, that trend stopped and began to reverse. Both trends are fully consistent with expected behavior of glaciers when considering the 206-year cycle and the RC theory.

A review of the US government's data of the status of the world's mountain glaciers confirms that glacial ice is now growing. Data from the Institute of Arctic and Alpine Research at the University of Colorado, Boulder, often working in conjunction with the National Snow and Ice Data Center, shows that although there has been a clear trend for the last hundred years or so in declining global glacial ice balance, that trend stopped around 1998–1999 and then began to reverse. While one can certainly go to selected glaciers and still show melting versus growth, the global trend has nonetheless changed. Again, this is in keeping with the RC theory predictions and the 206-year cycle's expected behavior.

The 2009 State of the Climate report issued by NOAA discusses the status of the world's glacial ice balance and many other climate indicators. However, this report fails to point out the change in the long-term trend of growth in global glacial ice and falsely implies a continued shrinkage that will be intensified without a reduction in greenhouse gasses. More importantly, since the report relies heavily on the same poor (or intentionally selected) data and bad interpretations of data used by IPCC, one should not view the report as in any way indicative of future climate trends. It has substantial other failings and erroneous conclusions, just like the IPCC reports. The report was an embarrassment to the many good scientists at NOAA and an indictment of the few bad ones and their leaders. This piece of pure political propaganda was intended to deceive Congress and the American people into voting in favor of the punitive, flawed climate change legislation proposed by the Obama administration. In part because of the blatant misrepresentation of climate science data in the 2009 State of the Climate report, I called for the firing of the president's science advisor and the NOAA administrator.

On the subject of when mountain glacial ice began its reversal from a declining to an increasing state, I have an important correlation. Using Swiss and US data, it can be seen that the reversal is intimately linked to the change in Earth's temperature profile and the end of global warming. This matches perfectly with the year 1998, which was the warmest this century, when the ice stopped shrinking. From that year on, there has been no effective growth in Earth's temperature, and the mountain glacial ice balance has begun to increase!

REASON 17:

We have been misled by the UN Intergovernmental Panel on Climate Change on the status of glacial ice melting.

The United Nations 2007 climate report issued by IPCC, AR4, contains fundamental errors, uses unreliable computer models and bad data, and outright lies to arrive at a set of profoundly false and misleading conclusions on the status of the world's glacial ice and most other future climate trends. For example, AR4 states that all Himalayan mountain glaciers would be gone by 2035, which has been shown to be a complete fabrication without scientific basis.[34, 35, 36, 37] Though an obvious and glaring mistake on the part of the UN and those trying to push the flawed AGW concept, it is far from the only one. The IPCC researchers did not discuss the issue with the experts on Himalayan glaciers in the government of India, who monitor the region's glaciers.[38] These genuine experts have stated there is no sign of abnormal retreat of Himalayan glaciers.

The issue of whether the world's glacial ice is melting or growing and why we have been force-fed only stories that it is melting warrants a much more detailed treatment than I can present here. There are many books and other references — including Internet sites hosted by real experts relying on well-researched data and arriving at objective, unbiased climate science conclusions — for any reader who wishes to get into a full treatment of the IPCC's attempts to mislead the world about what is happening with the Earth's climate, including the status of mountain glaciers. Earlier drafts of this book included sections on this subject that showed the misleading conclusions, bad data, and intentional deception in US government reports about the status of Earth's climate using data from the US government's own agencies. But again, for the purpose of this chapter, it is but one more solid indicator that our planet's climate is changing to a colder era and that we have purposely not been told the full story.

REASON 18:

There is a general lack of understanding among US and international science agencies about the real status of glacial ice.

My study of the available sources on the status of the world's glacial ice has revealed a disturbing fact: There is a wholesale lack of agreement within US science agencies, as well as among international sources, about just what the "official" total of the glacial ice is and whether it is growing or shrinking.

This statement about the overall reliability of our sum of knowledge about glacial ice points to the general weakness of any findings associated with them. My own conclusions were based on a data set that had to be cobbled together from multiple sources whose data appeared to be the most reliable, but definitely not anywhere close to being a gold standard. This sad situation forced me to create the Glacial Ice Data Standard (GIDS), shown in Table A3-2. It is an effort to identify how much glacial ice the planet contains and where it is located. This is the standard in use at the Space and Science Research Corporation; one day, hopefully, a global scientific community consensus on the subject can be established. Before we try to set an international standard, it would

TABLE A3-2. GLACIAL ICE DATA STANDARD. (GIDS)

1. **Totals:** If 100 percent of the world's glacial ice melted, sea levels would rise approximately 70 meters, or 230 feet.
2. **Antarctica:** It contains almost 90 percent of the world's glacial ice and, if completely melted, would raise sea levels 63 meters, or about 206 feet. It has an average thickness of 2,160 meters, or 7,086 feet.
3. **Greenland:** It contains approximately 8 percent of the world's glacial ice and, if completely melted, would raise sea levels by 5.6 meters, or about 18.4 feet. It has an average thickness of 2.3 kilometers, or about 7,546 feet.
4. **Mountain glaciers and other ice:** Mountains and other ice contain roughly 2 percent of the world's glacial ice and, if melted, would raise sea levels by 1.4 meters, or about 4.6 feet. Ice thickness has widely varying depths among the more than 160,000 mountain glaciers and other mountain ice concentrations.

be nice if we could start by getting the US science agencies to agree on one first.

REASON 19:

Destructive storms are not growing in number.

This incredible statement flies in the face of years of scare tactics from the AGW community. They have attempted to create a threatening menace called "man-made global warming," produced by CO_2 and other greenhouse gasses coming from our cars and trucks, utility power plants, heavy industry, chemical plants, farm animals, backyard barbecues, and last but not least, our own breath. According to the mantra of man-made ills, we are supposed to be experiencing, with each passing year of increasing CO_2, more cyclones, tropical storms, and especially hurricanes, which are expected to grow in number and intensity. The problem with this additional AGW myth is it has simply not happened. Again, the man-made climate change theory does not work. Here are elements of this proof:

In May 2007, the world experienced a record 33.4 days without a tropical cyclone on the planet. This means colder oceans.[39] This is a telling indicator of the state of the oceans around the world. And, as the oceans cool, so goes the rest of the planet. The last time a similar record occurred, it was during the 1980s minor solar minimum, which saw record-setting bitter winters, even in the southern US states. That particular cold affected many tens of thousands of acres of Florida's citrus groves, killing one third of Florida's commercial citrus trees.[40, 41]

A local TV meteorologist in Orlando, Florida, for NBC (WESH Channel 2), the ever-truthful Tony Mainolfi, in recognizing this unique bit of weather trivia news, had the courage to mention the new tropical cyclone record on air, saying, "So much for global warming!" I continue to see him with his typically thorough evening weather report, so apparently he didn't get fired for such a politically incorrect statement. There are still brave meteorologists out there and TV stations with the integrity and guts to back up their weather teams. Hopefully someday soon, all TV weather staff and their station managers will join the founder of The Weather Channel, John Coleman, me, and tens of thousands of weather and science professionals and publicly testify to what we all know: that the man-made climate change theory is the greatest science fraud of the modern era, if not all time. With the growing evidence of cold weather records being set, we may see more of them open up with their still-muzzled opinions. In the meantime, if I want the truth about my weather with an occasional honest comment about the climate, I will watch WESH TV and Tony Mainolfi.

Further, according to Table A3-3, we see that after the record Atlantic hurricane season of 2005 and up to 2009, hurricanes and all tropical cyclones have not grown in number and intensity around the world, as AGW advocates said they would. They are now at below historical average range, according to NOAA.

According to yet another source, despite what we have read in the papers, the number and intensity of the world's total of tropical cyclones does not seem to be on the increase. Dr. Phil Klotzbach at the Department of Atmospheric Science at Colorado

GLOBAL TROPICAL CYCLONES

Year	Number of Tropical Storms	NOAA Assessment
2006	78	Below Average
2007	79	Below Average
2008	90	Below Average
2009	82	Below Average

Source: SSRC. Data: NOAA

Table A3-3. Record of Global Tropical Cyclones from 2006 to 2009.

State University has studied global cyclone trends for the past 20 years (the peak years of Sun-caused global warming) and says for the period 1986 to 2005, "there has been no significant change on global net tropical cyclone activity."[42]

Or what about the results from a 2007 study[43] by the independent statistician, Dr. William Briggs, whose research paper stated the following:

> "We find that there is good evidence that the numbers of tropical cyclones over all ocean basins considered here have neither increased nor decreased since 1995: some oceans saw increases, others saw decreases or no changes."

In November 2006, there was a NOAA meeting of some of the world's best and brightest tropical storm experts in San Jose, Costa Rica. Here are some of their "Consensus Statements":[44]

- Though there is evidence for and against the existence of a detectable anthropogenic signal in the tropical cyclone record to date, no firm conclusion can be made at this point.
- No individual tropical cyclone can be directly attributed to climate change.
- The recent increase in societal impact from tropical cyclones has been largely caused by rising concentrations of population and infrastructure in coastal regions.

Why were these well-researched, concrete assessments by the experts not front page news or leading stories on the evening telecasts? Why, on the other hand, did we continue to see stories about the threat of worsening tropical storms each year the CO_2 count rose?

REASON 20:
The RC theory predicts another solar hibernation.

The theory of the Relational Cycles of Solar Activity predicts a solar hibernation based upon the same Bicentennial Cycle

THE NEXT SOLAR MINIMUM

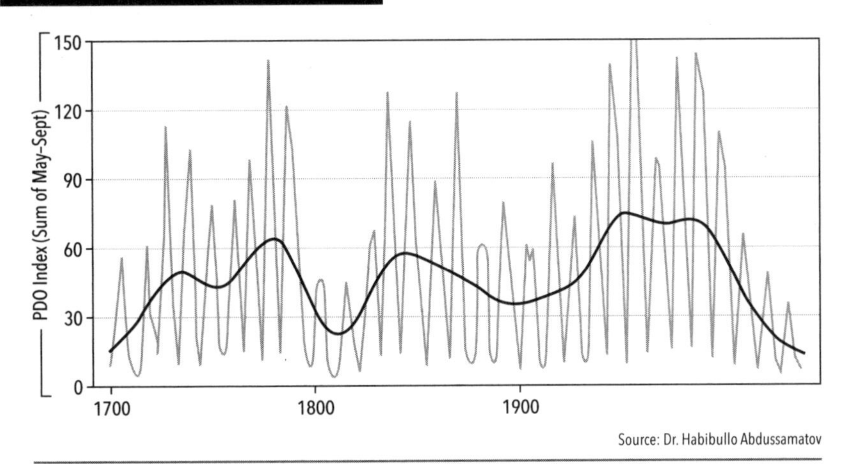

Figure A3-7. Projection of the next solar minimum. This chart displays, on the far right, the next three 11-year solar cycles as predicted by Dr. Abdussamatov, Russian Academy of Sciences. His estimates are for cycles showing sunspot numbers of low 70s for cycle 24, 50 for cycle 25, and then well below 50 for cycle 26, which is consistent with my own estimates of 74, 50, and 50.

happening in the same manner that it has done consistently for the past 1,200 years. Thus, it is going to happen again! No one can stop it. This is not a self-serving element of proof thrown into this appendix to create another reason. On the contrary, it is an essential element. There is a core need for a scientific theory to match observations — a pillar of the scientific method. The fact that one exists regardless of who developed it supports the contention that global warming has ended and a new cold climate has begun.

The Russians are once again way ahead of their US counterparts on this. Dr. Habibullo Abdussamatov at the Pulkovo Observatory has an almost identical prediction of the coming cold era to my 2007 forecast.

The development of a mathematical formula that reflects the performance of the Sun over the 206-year Bicentennial Cycle is important from a scientific standpoint. In the next two charts, we

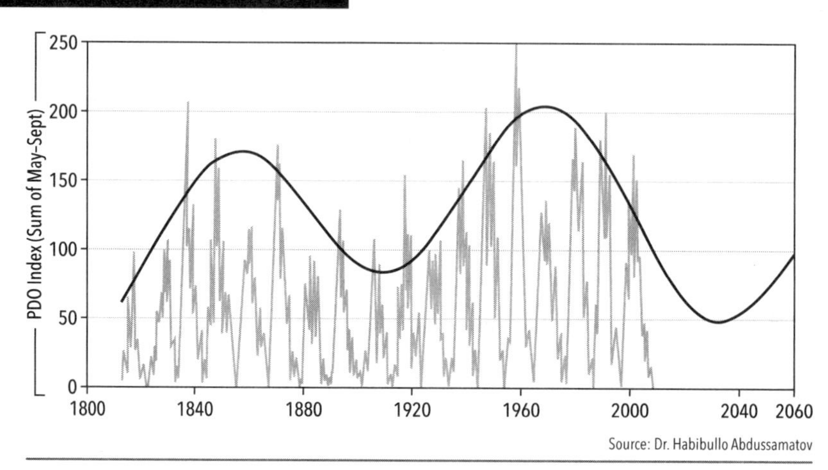

Figure A3-8. The 206-year Bicentennial Cycle over the last 200 years. Vukcevic's chart of the past cycles overlaid with his formula-derived curve for the 206-year cycle shows the coming solar hibernation low point at the far right. To the far left, around 1815, is the low point of the last solar hibernation during the Dalton Minimum. As of March 2011, we are headed down again toward the bottom in the 2030s right on schedule, as predicted. With it comes the cold.

see the mathematical formula in chart form for the 206-year Bicentennial Cycle based upon the independently derived formula by mathematician Milivoje Vukcevic.

It is important to restate a fundamental finding from my research into the development of the RC theory regarding the matching of sunspot curves and temperatures. While the peak of heating of the Earth based upon sunspot cycles is not coincident to high temperatures on the Earth, the cold weather low-temperature periods are strongly correlated. In other words, the sunspot bottom I have calculated to be in the year 2031 (agreed with by other scientists) and Vukcevic's formula above show a low long-term sunspot bottom that will be the same as Earth's cold temperature low. Note the mathematical sunspot peak extracted off Figure A3-8 at 1970, again the same as my previous calculation. Yet the peak of warming from the past Sun-induced global warming period was not 1970, but 2005–2007, which points to the temperature lag between sunspot maximums and temperature maximums.

Because of the high temperature incongruity with sunspot maximums, I decided that the RC theory would use the remarkably strong correlation with cold temperatures as its basis for prediction. This was a major decision in the theory's formulation. The RC theory is a cold Sun theory; it is a theory that tells us when the Sun will turn the Earth cold.

REASON 21:

Many scientists say a new cold climate is coming.

Prominent solar physicists and other scientists around the world tell us a new cold climate is coming. (See Appendix 1 for a detailed listing.) They have said the Sun is going into a historic, long solar minimum. We cannot in good conscience dismiss the research of so many respected scientists and researchers who have come to the same conclusion as I, that the Earth is headed for a pronounced extended cold climate period. A key feature in all these studies is they are not based on computer models with unproven assumptions but rather are strongly reliant on the Sun's past performance. Again, what we are all saying is that this has happened before, and since history and nature do repeat, it is happening again.

A fundamental point of the theoretical predictions by these highly experienced scientists is that they are no longer just predictions. These forecast events in general are now coming to pass. And they are already demonstrating real, measurable effects on Earth's climate. This compares in stark contrast to the man-made global warming concept, which has never shown itself to be accurate or reliable in predicting climate change, past or present, and therefore cannot be relied upon for future predictions. This has already been demonstrated by the concept's inability to predict the ongoing solar hibernation and new climate change to a long cold era.

REASON 22:

A rare conjunction of solar cycles is taking place.

In previous chapters, the RC theory was spelled out and its derivation explained. Element 7 of the theory proposes that there are other cycles, less than a hundred years in length, that may affect Earth's climate. Even in my early research, shorter cycles on the order of 50 to 60 years or less were evident. Research shows that along with the 206-year and the 90- to 100-year cycle in reversal, cycles of 50 to 60 years are also telling us a new cold climate is coming. One example covered above in the cooling oceans section is that of the Pacific Decadal Oscillation explained by Dr. Easterbrook. Another is a 50- to 60-year cycle of solar activity observed by several other scientists.

Of recent note is the work done by Dr. Nicola Scafetta in his April 2010 paper titled, "Empirical evidence for a celestial origin of the climate oscillations and its implications."[45] This outstanding and comprehensive paper has revealing findings and conclusions. For example:

- The paper cites evidence of three smaller solar cycles of 60, 20, and 9 years that influence Earth's climate.
- These cycles are related to the motion of the planets around the Sun and the moon around the Earth.
- Scafetta says the UN and greenhouse models fail to use solar cycles, which "outperform" current models.
- Climate sensitivity to man-made influences "has been severely overestimated by the IPCC by a large factor . . . Therefore the IPCC's projections for the twenty-first century are not credible."
- Scafetta concludes that over the next few decades, "global surface temperature will likely remain steady, or actually cool."

Dr. Scafetta's paper is one of the best I have read of literally hundreds on the subject of solar cycle influences on Earth's climate. He also gives due credit to the many scientists who have done extensive research that support his own findings.

This work by Dr. Scafetta and others, in conjunction with the RC theory, shows the important and unique time we live in when multiple, natural climate-determining cycles are coming together in a grand conjunction: the 206-year, 90- to 100-year, 60-year, 20-year, and 9-year cycles. There are likely other cycles that, with further research, could also be included.

There is little doubt in my mind that this grand conjunction of natural cycles marks one of the most influential periods in modern science, and in terms of how many lives may be affected, it may one day be regarded as one of the most important in all of the long history of science.

REASON 23:

Climate change research is poised to begin a new era of highly reliable forecasting based on solar activity.

I believe the solar physics and climate science communities have reached a watershed moment. The nature of climate change knowledge and prediction has reached a new and exciting stage in its evolution. Consider that:

- A substantial amount of research exists showing natural oscillations of the Sun have been the primary driver behind past climate changes.

- We now have an overwhelming amount of new, incontrovertible evidence showing solar cycles have been the most important agents of recent climate shifts of the past 200 years, including the most recent period of global warming.

- Climate models using solar activity are the only ones able to explain the now ongoing shift to a new long-term cold climate era.

- The solar cycle–based models for predicting future climate changes, compared to others, have been shown to be the most reliable by a wide margin.

In the absence of any other viable climate change model, I believe there is now a sufficient body of evidence to recommend that mankind's attempts to more accurately predict future climates on Earth should be primarily based on solar activity cycles, which are caused by the Sun's internal processes acting in concert with the other planets and our Earth-Moon system.

A solar hibernation has begun in our lifetime.

The most powerful in the RC family of cycles is, of course, the 206-year Bicentennial Cycle, which creates a solar hibernation. My prediction of a change in direction of the 206-year cycle is the catalyst for my research, the formation of the SSRC, and my drive to alert people around the world to the nature of the next climate era. In this section, you will read everything you need to know demonstrating that the next solar hibernation has started.

For the first time in over 200 years, we are living through the start of one of the most amazing astronomical occurrences in the history of science: a solar hibernation. Like before, it will bring decades of destructive cold weather, record earthquakes, and volcanic eruptions. Equally remarkable — and despicable — is the fact that you are being kept in the dark about it!

Here then are the reasons why you should believe in the first solar hibernation of the twenty-first century. If one can accept that the hibernation has in fact begun, just as I predicted it would in April 2007, then the following elements of proof should be sufficient for every citizen on the Earth to prepare for the coming new cold climate era:

REASON 24:

NASA says a major solar minimum (solar hibernation) is coming![46, 47]

At this point, we can all close this book and go to the nearest L. L. Bean and buy more winter clothing! When I started the process of passing on what I had detected in the signature of the sunspot curve, NASA was among the first groups I contacted. They had already come to their own conclusion that the Sun was doing something unusual and had concluded that there would indeed be a steep drop-off in sunspot count, yet it would not begin until solar cycle 25, the one after the one that just started, cycle 24, and not for another 11 years. This information from NASA, even in 2007, was universally accepted within the astronomy and solar physics community but was once again kept from the general public except for the few students of solar activity with the right amount of interest and knowledge to pull it off the Internet. Coming with this minimum, as I have shown, will be a period of intense cold — a record-setting deep cold. NASA and NOAA do not yet accept the term "solar hibernation." The scientific community prefers to call such rare solar events a "grand minimum." I came up with the term "solar hibernation" after trying to explain to friends what it was all about. The new term proved to be far easier to comprehend and is a much better description of what the Sun is doing.

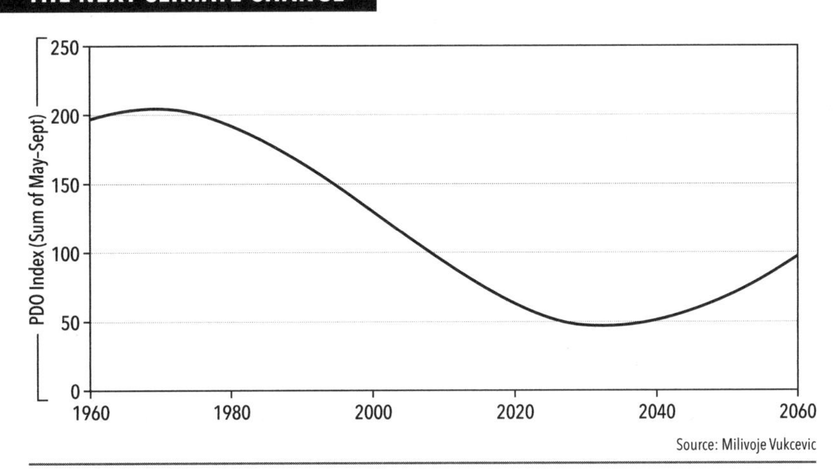

Figure A3-9. The low point of the next climate change. Vukcevic's formula shows identical outcomes to my own calculations of the next 206-year cycle with a minimum in the year 2031. Again, this is a sunspot curve, not a temperature curve. However, global temperatures at their coldest and the bottom of the sunspot curve do coincide.

Still the message, even in 2006, was clear: NASA had concluded that the Sun was about to undergo a significant change, one that had not been seen for hundreds of years. When I made my initial call to the Solar Physics Group at NASA's Marshall Space Flight Center in early 2007, the group leader, Dr. David Hathaway, was kind enough to listen as I bluntly told him that NASA's predictions for the approaching solar minimum were "way off."

At that time, they were forecasting a near record peak of solar activity for cycle 24, with sunspot counts of 145 or more and a peak in the year 2011. I extended my matter-of-fact opinion to Dr. Hathaway, saying that according to my calculations, the next cycle, 24, would peak in 2012 and that the sunspot count would not be greater than 74 — half what NASA was saying!

It's important to understand the context of this conversation with one of NASA's top solar physicists. Here I was, with not one scientific paper to my name, no record of research in the field, and no academic standing, telling NASA that they had made a major miscalculation in what the next solar cycles were going to look like. Hathaway could have slammed the phone down, but he was too professional to act that way. Instead he said that yes, NASA and NOAA scientists were revising the predicted peak year but he would not concede on the sunspot count. They were going to stick by their past prediction that cycle 24 was still going to be one of the most active in the history of solar cycle record keeping.

NASA and NOAA scientists regularly get together to review the latest data on solar activity, and after meeting on the topic, they update their forecast. I tried to pass on my thinking to key officials and scientists at NOAA and later through the NOAA administrator's office, but to no avail. Since my first announcement that the next solar hibernation would begin in cycle 24, with a record low sunspot count for both cycle 24 and 25, I had been relegated to the category of dubious researchers, included with a handful of other established scientists around the world who had come to the same conclusion yet whose opinion had no merit.

Our data and findings were at the end of the bell-shaped curve, beyond the two sigma line and far from the official forecast by the conventional scientific community. We were of course guilty of a more grievous sin than not conforming to the conventional scientific opinion: We were politically incorrect.

So what has happened since those first contacts almost four years ago with both NASA's Hathaway and his counterparts at NOAA? Not surprisingly, both agencies, containing our country's best government solar physicists, have had to adopt a program of continual revision of their forecasts for solar activity. Their predictions came and went as the years passed and routine conferences between them had to deal with a rapidly changing "space weather" environment. The reality of the Sun's behavior and the ever more obvious signs of the next solar hibernation have forced both groups to routinely downgrade their sunspot projections and slip the cycle 24 peak further right, to where today it is almost exactly identical with mine and that of the other colleagues who were bold enough to go against the grain back then. The January 2011 announcement by NASA for the current solar cycle 24 was a momentous event that, to the extent of what it means to you and me and everyone else on the planet, was totally missed by every major media outlet that I'm aware of. It was not missed, however, by those who follow the Sun and know what this fundamental change in the Sun means to Earth's climate. I issued the following press release to try to make it clear to the media and others just how this NASA announcement affects us all:

Press Release 1-2011 (edited)

NASA Data Confirms Solar Hibernation and Climate Change to Cold Era

Tuesday, January 25, 2011

3:00 PM

The Space and Science Research Center (SSRC) announces today that the most recent data from NASA describing the unusual behavior of the Sun validates a nearly four-year-long quest by SSRC Director John L. Casey to convince the US government, the media, and the public that we are heading into a new cold climate era with 20 to 30 years of record-setting cold weather.

According to Director Casey, "I'm quite pleased that NASA has finally agreed with my predictions which were passed on to them in early 2007. There is no remaining doubt that the hibernation of the Sun, what solar physicists call a 'grand minimum,' has begun and with it, the next climate change to a prolonged cold era.

"When I first called Dr. Hathaway and told him the NASA and NOAA estimates for the Sun's activity were 'way off' in both sunspot count and in which solar cycle the hibernation would begin (cycle 24 vs. cycle 25), he was polite but dismissive. Since that time, both NASA and NOAA have been revising their sunspot estimates for solar cycle 24 lower every year and with each year their numbers have been getting closer to mine and the few other scientists around the world who had similar forecasts. The January 2011 announcement by NASA is now virtually identical to mine made almost four years ago."

NASA's Solar Physics Group, headed by Dr. David Hathaway at the Marshall Space Flight Center, alerted the solar physics community on January 3, 2011, that the latest sunspot prediction for our current solar cycle 24 had been adjusted downward by a significant amount from recent years to a value of 70 ± 18 and an estimated peak of 59 sunspots during solar maximum in the June-July 2013 time frame. This number compares with their prediction of a much larger 2006 estimate of a very active Sun with 145 sunspots at peak. Many of the gauges by which the Sun's activity is measured, like sunspot counts, have since set record low levels. Casey's 2007 forecast, however, came during the height of the man-made global warming movement at a time when any mention of a reduction in the Sun's energy output, much less a new cold climate, was political and scientific heresy.

As Casey recounts, "Once I made my forecast for the Sun's reversal in phase from global warming to global cooling and the start of a new cold climate period, I was immediately attacked from all sides. Regrettably, that is the history of new scientific discoveries when anyone says the opposite of a belief that is well entrenched in conventional thinking. My prediction also ran into political roadblocks since at that time both presidential candidates were trying to woo the 'green' vote in what all knew was going to be a close election where every vote counted. Both Republicans and Democrats were saying man-made global warming was real and something should be done about it. Despite my strong space program credentials, what I was saying then was a message no one wanted to hear. Both liberal and conservative websites launched attacks to discredit my research. Fortunately, the Sun has been on my side and it is a powerful ally. At long last, NASA has now come out with their own data that confirms my past predictions.

"After I had completed my original research and notified NASA, I tried to find others who had come to the same conclusion about the Sun and the next climate change. I want to take the time today to mention some of these prominent researchers who made the courageous step forward back then and went public with their predictions. The list is also posted at the SSRC website. They include in the US: Drs. Ken Schatten, D. V. Hoyt, and W. K. Tobiska; in Europe and Russia: Drs. Habibullo Abdussamatov, Oleg Sorokhtin, Boris Komitov, Vladimir Kaftan, O. G. Badalyan, V. N. Obridko, J.

Sykora, and J. Beer; in Australia: David Archibald, Drs. Ian Wilson, I. A. Waite, Bob Carter, and Peter Harris; in China: Drs. Y. T. Hong, H. B. Jiang, L. P. Zhou, H. D. Li, X. T. Leng, B. Hong, X. G. Qin, L. Zhen-Shan, and Sun Xian; and in Mexico: Dr. Victor M. V. Herrera. I also want to express my thanks to and hope to soon add the many more researchers to this partial list who have supported the position that the Sun drives climate change, not mankind, and that we have begun the transition to the next cold climate."

As to the linkage of the new cold era with this now confirmed solar hibernation by NASA, Director Casey clarified, "NASA is not the primary source for US government weather and climate forecasts. With the exception of NASA Goddard, that's NOAA's area of responsibility, though we all rely on the data from weather satellites that NASA launches into orbit around the Earth and the Sun. But don't ask any of the NASA or NOAA scientists to agree with the end of global warming and the now confirmed start of the next solar hibernation or for that matter a cold climate change. That would be career suicide given the measures the current administration goes to in order to preserve the myth of man-made global warming. In any case, decades of extreme cold weather always follow these hibernations of the Sun as the research shows going back 1,200 years or more. This next one has begun right on schedule, just as I predicted. We should therefore expect the same climate change to a long cold period just like it has done before. The last three record cold and long winters around the globe, along with the lack of growth in the planet's average temperature for the past 12 years, and a new long-term downward trend in global temperatures are solid enough signals to prove that global warming ended as and when I predicted and that the Earth is rapidly proceeding into a long cold era.

"NASA's announcement is clearly vindication for those of us who have spoken out for years against conventional climate science thinking, false statements and misleading reports of the UN and US government climate science officials, and had to endure slander and ridicule from AGW extremists. Now we need to prepare for what has arrived: 20 to 30 years of record-setting, crop-destroying cold weather. We should stop wasting precious resources on the past climate phase of Sun-caused global warming, bury this hubris of man-made climate change, and listen to what the Sun is telling us. We need to do so immediately."

See the new NASA Solar Cycle 24 prediction at:
http://solarscience.msfc.nasa.gov/predict.shtml

This critical NASA data from January of 2011 and the associated SSRC press release provide for you the essence of the solar hibernation prediction and its confirmation. The measurements of the sunspots on the Sun are not the only indicator of a dropping solar energy output. The remaining reasons for believing in the next climate change follow.

REASON 25:

The surface movement on the Sun "has slowed to a record crawl."[46]

This information about one of the most important indicators of the Sun's activity level also comes from NASA and has been posted on their website since at least May 10, 2006. They say the Sun's outer shell, its sunspot-creating exterior, is "off the bottom of the charts" in terms of reduced movement, hence reduced solar output and a colder Earth-Sun environment. As a result, in the words of NASA's Dr. Hathaway, "Solar Cycle 25,

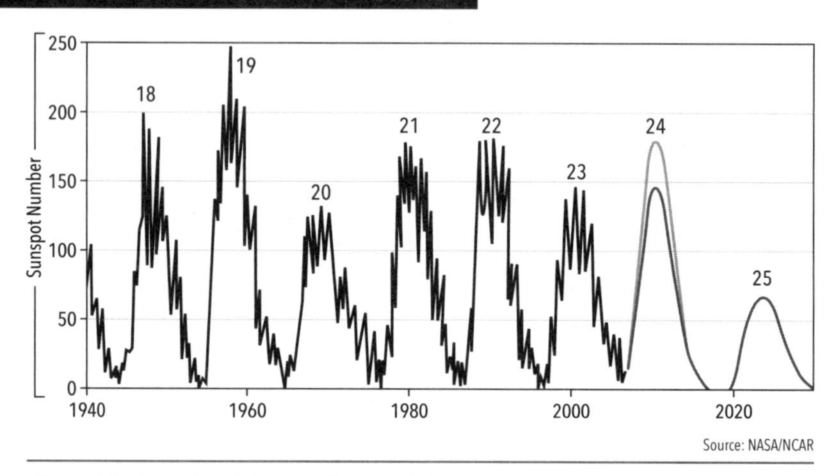

Figure A3-10. NASA's 2006 prediction of the next two solar cycles. This 2006 chart from the Solar Physics Group at NASA's Marshall Space Flight Center depicts their opinion that cycle 24 would be another strong one, with an average count of 145 sunspots. It also shows that even then, NASA said a major solar minimum would occur in cycle 25 (albeit 11 years off the mark).

peaking around the year 2022, could be one of the weakest in centuries."

The solar hibernation and its implications should not only be breaking news on evening TV stations and in newspaper headlines, but should also be the hottest current-event subject in every college, high school, and elementary class throughout the United States! The solar hibernation is one of the most astounding events in the history of science, with worldwide impacts on every individual, and yet it has been virtually buried by those who control how critical scientific information is communicated to the public.

REASON 26:
The solar wind is at a 50-year low.[48]

There are a number of ways to chart the activity of the Sun. One of these is to measure the constant blast of superheated high-energy particles that are ejected from the Sun each second.

This "solar wind" clears a path and creates an envelope surrounding our Earth and the entire solar system. On September 23, 2008, NASA held a media teleconference to report that the joint NASA and European Space Agency Ulysses satellite mission had determined that a 50-year low point in the solar wind had been detected. Again, this is conclusive evidence that the Sun is heading into hibernation. The Sun's output is shutting down to a level not seen for generations, and the process is not over yet.

REASON 27:

The planetary magnetic field strength (Ap) is at an all-time low.

The Earth's planetary magnetic field in its interaction with the Sun is measured at 13 stations around the world. Data is integrated at the GeoForschungsZentrum in Potsdam, Germany, and distributed to scientists around the globe (prior to the year 1997, such data was provided by the Institut für Geophysik in Gottingen, Germany). The Ap index is now at an all-time low reading, as shown in Figure A3-11.

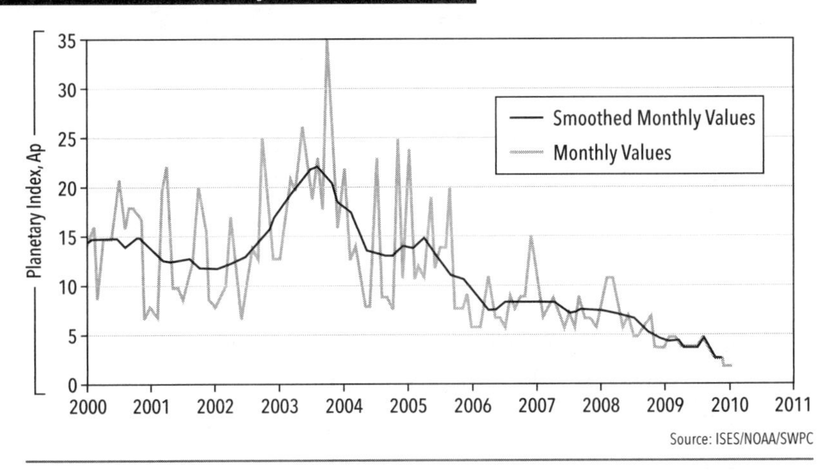

Figure A3-11. Planetary magnetic field index, Ap. This one-year duration chart of the Ap index, current through February 2, 2010, is a measure of Earth's magnetic field as it interacts with that of the Sun and shows a record decline is in place. This is another significant confirmation of the presence of a solar hibernation.

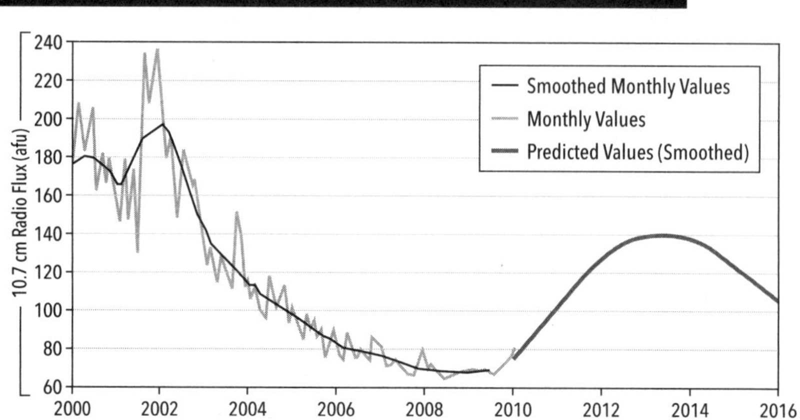

Figure A3-12. Measurement of solar activity based on the 10.7 cm radio wavelength. The measurement of the 10.7 cm wavelength shows that for the past three years, it has stayed in the 65–80 range and has done so longer than any time previously measured.

REASON 28:

The Sun's radio flux density is at an all-time low since record keeping began in the 1950s.[49]

Another measurement of the activity of the Sun is its output of energy in the 10.7 cm wavelength in the radio wave portion of the solar spectrum. The chart in Figure 5-12 displays the area between cycles 23 and 24 and, like sunspot activity, shows that it has been at the bottom for the better part of three years.

REASON 29:

Cosmic rays from outside the solar system have reached the highest level ever recorded, indicating the Sun's protective envelope around the Earth has never been weaker.

A weakened solar wind will in theory allow cosmic rays from outside our solar system to enter our atmosphere. It is a central tenant to the work of Dr. Henrik Svensmark in showing cosmic

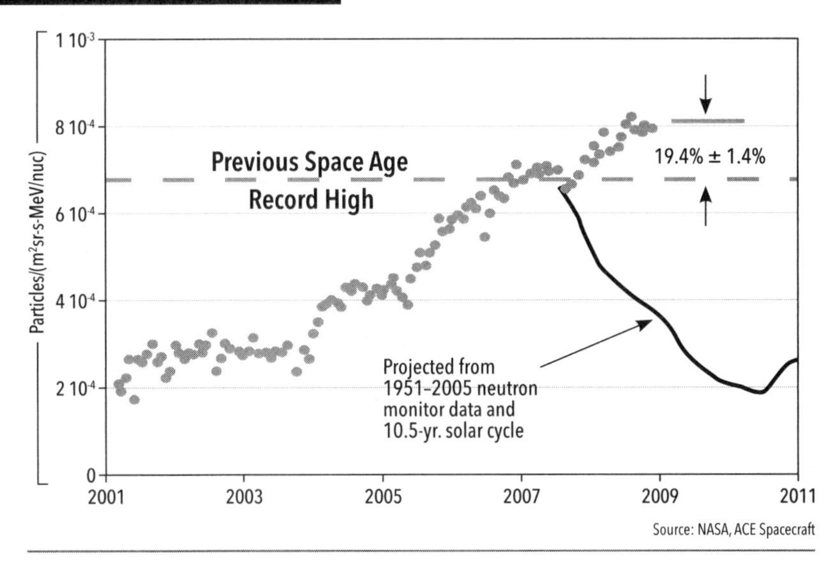

Figure A3-13. Cosmic ray iron (Fe) nuclei at highest level ever recorded. The more cosmic rays we receive, the more iron nuclei we count. This is another solid confirmation of a solar hibernation, as a weaker Sun produces a weaker solar wind, one less able to protect us from cosmic rays. The curve to the bottom right of the chart shows how the solar hibernation has once again dashed NASA's projections for solar activity. *270–450 MeV/nucleon.

rays develop greater cloud cover and are a possible mechanism for cold weather production.

David Archibald has also documented this record increase in the influx of cosmic rays using data from the University of Oulu, Finland.[50]

A second source for cosmic ray measurement comes from NASA.[51] A NASA plot of cosmic rays in the form of iron nuclei is shown in Figure A3-13. In this figure, we see that cosmic rays measured in the form of iron (Fe) isotopes have jumped 19.4 percent above previous records going back to when NASA started space flight, establishing a new record level.

REASON 30:
The Sun is shrinking!

Let me say that again: The Sun is shrinking! This is according to Russian physicists using leading-edge technology, who have done extensive measurements and have tracked the reducing solar diameter for years. This solar behavior is similar to a hot air balloon that expands as the gas jets in the basket heat up the interior envelope of the balloon. Without continued heating, it will contract. The shrinking of the Sun may possibly be a result of a lowered solar energy output, hence a colder Earth-Sun environment. The root causes and energy transfer mechanisms within the Sun are not well understood and still require much research. According to Dr. Habibullo Abdussamatov of the Russian Academy of Sciences, the Sun's radius has been shrinking since 1979 and by 2018 will have been reduced by 81.6 kilometers. While that may not seem a large number when compared with the Sun's 695,990 kilometer radius, for an energy generator the size of the Sun, that amount of reduction could indicate a substantial energy

THE SHRINKING SUN

Cycle	$S_\circ$, W/m²		$S_\circ^{max} - S_\circ^{min}$, W/m²	W_{max}	▲R_m, km
	maximum	minimum			
21	1366.49±0.06 (12/1979)	1365.57±0.03 (08/1986)		164.5 (12/1979)	0 (08/1986)
22	1366.46±0.06 (07/1989)	1365.55±0.03 (05/1996)	0.89	158.1 (07/1989)	-5.1 (05/1996)
23	1366.41±0.06 (04/2000)	1365.43±0.03 (07/2007)	0.86	120.7 (04/2000)	-35.7 (07/2007)?
24	1366.24±0.15? (07/2011)	1365.25±0.10? (09/2018)	0.81?	70.0±10.0? (07/2011)	-81.6? (09/2018)?

Source: Dr. Habibullo Abdussamatov, August 28, 2006, Solar Physics Vol. 23, No. 3, 91–100

Table A3-4. Note cycle 24 and then go to the far right to see the estimated 81.6 km shrinkage by 2018.

reduction. Table A3-4 shows Abdussamatov's measurements of the Sun's shrinkage.

Has anyone read this in their local paper or heard that the Sun is shrinking on the evening news? Those who control what you read and hear have done an excellent job making sure you do not hear about this one.

Dr. Abdussamatov, one of Russia's best scientists, is another researcher who has predicted a long-term cold period.

REASON 31:

The northern lights are at a record low.

The aurora borealis, or northern lights, have fascinated mankind since first observed. These lights, visible as waving curtains of multicolored lights in the sky in the northern latitudes, are created by the Sun's high-speed rays interacting with Earth's magnetic field. They tend to follow the 11-year solar cycle in terms of waxing and waning and have been used as an indicator of the Sun's output. On September 28, 2010, the Finnish Meteorological Institute said that the aurora had declined to a hundred-year low. Clearly the Sun has weakened to historical levels.[52]

REASON 32:

The Earth's upper atmosphere is shrinking.

One would expect that if the Sun is putting out less energy, then Earth's atmosphere would exhibit that in several ways. One way is to shrink, like any gaseous area will do around an orbiting planet. And that is exactly what has been happening. Scientists in two recent reports, in June 2010 and August 2010, have found that the thermosphere, Earth's superhot top layer, had shrunk significantly. In the first report, lead study author John Emmert at the Naval Research Laboratory said it was the "biggest contraction in the thermosphere in at least 43 years."[53] In the second study, by the National Center for Atmospheric Research in Boulder, Colorado, it was announced that the thermosphere shrank by

30 percent because of a sharp drop in ultraviolet radiation from the Sun. According to the lead author of the study, Dr. Stanley Solomon, and co-author Dr. Thomas Woods, it contracted "more than at any time in the 43-year era of space exploration." Woods added, "If it is indeed similar to certain patterns in the past, then we expect to have low solar cycles for the next 10 to 30 years."[54] Lower solar cycles for the next 10 to 30 years? Here again, not only do we see another result of the declining Sun, but an indirect reference to the predicted solar hibernation and its 20 to 30 years of reduced solar output.

REASON 33:
Solar irradiance is declining.

At the heart of the deception found in the IPCC reports that attempts to convince us that the Sun plays only a small role in climate change is the allegation that measurements of the Sun's energy delivered to the Earth's surface show it varies by too small a percentage to have any effect. Thus they dismissed the Sun as having any strong "forcing" function on Earth's climate. The evidence they and the AGW crowd display are the data and charts of the energy calculated in watts per meter squared that the Sun delivers over time. In essence, these charts show the Sun's energy, called "solar irradiance," changes in sync with the 11-year solar cycle and with an average of 1,366 watts per meter squared. They say that the typical variation in this number, about 1.3 watts per meter squared, is too small to affect the Earth's climate. And that, my friends, is where the fault lies.[55]

To put this in perspective to show just how much energy the Sun delivers every day, just compare this to a typical microwave in a kitchen. Many are rated to run at 1,200 watts energy output. We all know how hot a microwave can get. Now imagine a microwave running constantly for every square meter of your house, your garage, your front and back yard, your street, your neighborhood, and on and on. Our great Sun does this for us every day! We have no energy shortage — what we have is a shortage of the will and means to capture the free energy the Sun already provides for us.

What we now know is that even the smallest changes in the intensity of the Sun can have significant effects for us, and whether the Earth will be getting colder or warmer, for decades at a time. There are many studies that have been done in this area to understand the solar forcing function. One that seems to put it all in perspective was done by Dr. Bas Van Geel et al. and published in 1999, the year after the record global temperature was set.[56] Here is one of his team's critical conclusions:

"The climate system is far more sensitive to small variations in solar activity than generally believed. For instance, it could mean that the global temperature fluctuations during the last decades are partly, or completely explained by small changes in solar radiation."

This group of scientists is not alone in their assessment of how the Sun is the major force behind climate change. I have read many similar research studies. This singular item is included here to show once again that there is much about the Sun we have not been told.

So what is the status of the Sun's output as measured by its irradiance? A study of the total solar irradiance (TSI) measured by several satellites shows reason to believe that the solar hibernation has begun. The last three decades during which we have had access to TSI measurements via satellites show that the Sun's irradiance is at the lowest levels in 12 years and is close to the lowest levels since measurements began in the mid-1970s! Go to "Deep Solar Minimum — NASA Science" at science.nasa.gov to see their chart of the Sun's TSI output.

This appendix has set forth 33 reasons for believing a new cold climate has begun; these reasons are built upon rock-solid evidence. The predicted solar hibernation and the cold climate era that it will bring have now arrived. We must therefore begin to prepare.

The first step in that preparation is understanding the truth about what the primary cause of climate change is: namely, the Sun, not humans.

The second step is the really tough part. We must accept that in this age of advanced communications, we — you and I, and the rest of the world — have been misled and deceived on a global scale for many years.

The third step comes in the form of a question. The question I pose to you is straightforward. Now that you have this knowledge, this truth, will you fear it and do nothing, or, like Thomas Jefferson might recommend, will you help me in telling it "to the whole world"?

Notes

Chapter 1: Moment of Revelation . . . Wow!

1. NOAA, National Climate Data Center. "Climate of 2007 April in Historical Perspective."

2. Ashley Gosik. (2007). "Silver tsunami may test Social Security system." *Orlando Sentinel*, October 13. Cox News Service.

3. Wikipedia. "Year Without a Summer." http:/en.wikipedia.org/wiki/Year_Without_a_Summer.

4. NASA/MSFC. "The Sunspot Cycle." May 3, 2007.

5. galileo.rice.edu/sci/observations/sunspots.html

6. Wikipedia. "Holocene."

7. Wikipedia. "Solar Variation."

Chapter 2: What Happened the Last Time?

1. Wikipedia. "John Dalton."
2. I. G. Usoskin, K. Mursula, and G. A. Kovaltsov (2002). "Lost sunspot cycle in the beginning of Dalton Minimum: New evidence and consequences." Geophys. Res. Lett., 29(24), 2183,doi:10.1029/2002GL015640.
3. Population Reference Bureau. (2007). "World Population Data Sheet." http://www.prb.org/Publications/Datasheets/2009/2009wpds.aspx.
4. Population Reference Bureau. (2005). "World Population Growth, in Billions" (chart).
5. US Census Bureau, International Database. "Total Population of the World by Decades, 1950–2050."
6. US Census Bureau. "1800 Census." http://1930census.com
7. About.com. "U.S. History: Presidents and Vice-Presidents of the United States."
8. Adam Zamoyski. (2004). "Out in the cold. A review of 'Moscow 1812: Napoleon's fatal march." *Washington Post*, August 8.
9. The Louisiana State Museum.
10. Robert Evans. "Blast from the Past." Smithsonian.com.
11. Wikipedia. "Year Without a Summer."
12. Willie Soon and Steven H. Yaskell (2003). "Year without a summer." *Mercury*, May–June.
13. Jelle Z.de Boer and Donald T. Sanders (2002). *Volcanoes in Human History*. Princeton, NJ: Princeton University Press.
14. USGS. "West Indies Volcanoes," vulcan.wr.usgs.gov/Volcanoes/WestIndies/description_west_indies_volcanoes.html>

15. Wikipedia. "Mayon Volcano," http://en.wikipedia.org/wiki/Mayon_Volcano
16. J. Neumann (1989). "The 1810s in the Baltic region, 1816 in particular: Air temperatures, grain supply and mortality." Springer Netherlands, Vol. 17, No. 1, August 1990.
17. David Archibald. "The Past and Future Climate." May 2007. From a public presentation by Archibald.
18. Wikipedia. "Laki." en.wikipedia.org/wiki/Laki
19. Jill Lawless. (2010). "Iceland's volcanic ash halts flights across Europe." Associated Press, April 15.
20. "Iceland's Eyjafjallajokull volcano is nothing to 'Angry Sister' Katla." *Christian Science Monitor,* April 18, 2010, www.csmonitor.com/Science/2010/0418/
21. "Volcano erupts again." *Orlando Sentinel,* November 13, 2010. From News Services — no author cited.
22. USGS. http://earthquake.usgs.gov/regional/states/10_largest_us.php.
23. USGS. "M8.9 Near the East Coast of Honshu, Japan." earthquake.usgs.gov/pager
24. USGS. "USGS Updates Magnitude of Japan's Tohoku Earthquake to 9.0." www.usgs.gov/newsroom/article.asp?ID=2727&from=rss_home
25. *Encyclopedia Britannica Online.* "Parisian revolt." December 2, 2007.

Chapter 4: The Future

1. Population Reference Bureau and United Nations. (1998). "World population projections to 2100."
2. P. T. Higgins, "Climate change in the FY 2011 budget." American Meteorological Society.

Appendix 1: Scholarly Acknowledgments

1. David H. Hathaway and Robert M. Wilson (2004). "What the sunspot record tells us about space climate." *Solar Physics* 224:5–19.

Appendix 3: Press Releases — Global Warming Has Ended; The Next Climate Change Has Begun

1. Hadley Centre for Climate Research (HadCRUT) and Anthony Watts. "Global temperature anomaly through January 2008." www.wattsupwiththat.com
2. University of Alabama, Huntsville, and Anthony Watts. "Monthly means of lower troposphere." www.wattsupwiththat.com
3. Remote Sensing Systems (RSS) and Anthony Watts. "Monthly means of the lower troposphere, V3.01, January 1979 to May 2008." www.wattsupwiththat.com
4. National Climate Data Center. "Significant climate anomalies and events in 2007." December 13, 2007. www.ncdc.noaa.gov/oa/climate/research/2007/ann/ann07.html
5. M. Tolson, E. Berger, M. Glenn, and C. George (2009). "Snow drops in for a visit." *Houston Chronicle*. December 5. chron.com/disp/story.mpl/6750042.html
6. E. Berger, C. Horswell, and M. Tolson (2010). "Batten down for a frigid blast." *Houston Chronicle*. January 6.
7. "Europe and ASIA struggle with travel, power woes." *Wall Street Journal*. January 6, 2010.
8. J. Ruwitch (2008). "China battles 'coldest winter in 100 years.'" Reuters, February 4.
9. P. Vinthagen Simpson (2010). "Sweden braces for record freeze." *The Local*, November 30. www.thelocal.se/30516/20101130/

10. “Coldest winter in 1,000 years on its way.” RT, Prime Time Russia, October 4, 2010. www.rt.com/news/prime-time/coldest-winter-emergency-measures

11. Anthony Watts and Weather Underground. “‘Gore effect’ on steroids: Six straight days of record low temperatures during COP16 in Cancun Mexico — more coming.” December 10, 2010.http://wattsupwiththat.com/2010/12/10gore-effect.... mexico.

12. J. Broder (2010). “Climate talks end with modest deal on emissions.” *New York Times*. December 11.

13. A. Doyle (2010). “Expectations mild for global warming talks.” *Orlando Sentinel*, November 25 (Reuters).

14. “December was coldest in 120 years.” Telegraph.co.uk, January 4, 2011.

15. “Britain could be heading for coldest winter in 300 years.” Telegraph.co.uk, January 4, 2011.

16. J. M. Lyman, J. K. Willis, and G. C. Johnson (2006). “Recent cooling of the upper ocean.” *Geophysical Research Letters*, Vol. 33. September 20. L18604. doi:10.1029/2006GL027033,2006

17. J. K. Willis, J. M. Lyman, and G. C. Johnson (2007). “Correction to ‘Recent cooling of the upper ocean.’” Submitted to *Geophysical Research Letters*. June 8.

18. D. Avery (2005). “California seal pups predict Pacific Ocean cooling.” National Center for Policy Analysis. June 20.

19. Anthony Watts, “Shifting of the Pacific Decadal Oscillation from its warm mode to cool mode assures global cooling for the next three decades.” July 20, 2008, http://wattsupwiththat.com.

20. NASA April 21, 2008. "Larger Pacific Climate Event Helps Current La Niña Linger." http://www.jpl.nasa.gov/news.cfm?release=2008-066.

21. NOAA January 2011. "2010 Wet Season Update." http://www.wrh.noaa.gov/mfr/cliom/update_lanina_2011.php.

22. Peter T. Doran et al. (2002). "Antarctic climate cooling and terrestrial ecosystem response." *Nature* 415, January 31, 517–520. doi:10.1038/nature710

23. C. H. Davis, Y. Li, J. R. McConnell, M. M. Frey, and E. Hanna (2005). "Snowfall-Driven Growth in East Antarctic Ice Sheet Mitigates Recent Sea Level Rise." *Science*, doi:10.1126/science.1110662.

24. IPCC. (2007). "Summary for policymakers." In *Climate Change 2007: The physical science basis. Contribution of Working Group 1 to the Fourth Assessment Report of the Intergovernmental Panel on Climate Change* (Solomon, S., D. Qin, M. Manning, Z. Chen, M. Marquis, K. B. Averyt, M. Tignor, and H. L. Miller, eds.), Cambridge, U.K., and New York: Cambridge University Press, p. 17.

25. M. Tedesco and A. J. Monaghan (2009). "An updated Antarctic melt record through 2009 and its linkages to high-latitude and tropical climate variability." *Geophysical Research Letters*, 36, L18502, doi:10.1029/2009GL039186.

26. E. Hanna and J. Cappelen (2003). "Recent cooling in coastal southern and relation with the North Atlantic Oscillation." *Geophysical Research Letters* 30: 10.1029/2002GLO15797.

27. B. M. Vinther, K. K. Andersen, P. D. Jones, K. R. Briffa, and J. Cappelen (2006). "Extending Greenland temperature records into the late eighteenth century." *Journal of Geophysical Research*, 111, D11105, doi:10.1029/2005JD006810.

28. H. Jay Zwally, Mario B. Giovinetto, Jun Li, Helen G. Cornejo, Matthew A. Beckley, Anita C. Brenner, Jack L. Saba, and

Donghui Yi. (2005). "Mass changes of the Greenland and Antarctic ice sheets and shelves and contributions to sea level rise: 1992–2002." *Glaciology*, Vol. 51, No. 175, December, pp. 509–527 (19).

29. Ola M. Johannessen, Kirill Khvorostovsky, Martin W. Miles, and Leonid P. Bobylev (2005). "Recent ice-sheet growth in the interior of Greenland." *Science*, Vol. 310, November 11, pp. 1013–1016.
30. Arctic Regional Ocean Observing System (Arctic-ROOS), Nansen Environmental and Remote Sensing Center. arctic-roos.org/observations/satellite-data/sea-ice/ice-area-extent-in-arctic
31. Anthony Watts. February 2009, "George Will's battle with hotheaded ice alarmists." www.wattsupwiththat.com
32. Animal Info Endangered Animals. *Polar Bear*. December 18, 2010. www.animalinfo.org
33. IUCN, Polar Bear Specialist Group. "Summary of polar bear population status per 2010." March 2010, pbsg-npolar.no/en/status/status-table.html
34. IPCC. (2007). AR4. Chapter 10, p. 493.
35. IPCC. (2010). "IPCC statement on the melting of Himalayan glaciers." January 20.
36. World Wildlife Fund. "Correction. Acknowledgment of error in quoting melting of Himalayan glaciers by 2035." Insert to original paper "An Overview of glaciers, glacier retreat, and subsequent impacts in Nepal, India, and China." March 2005
37. World Wildlife Fund (2005). "An overview of glaciers, glacier retreat, and subsequent impacts in Nepal, India, and China." March 2005.
38. F. Pearce (2010). *New Scientist*. "Debate heats up over IPCC melting glaciers claim." January 13.

39. "Longest period without a tropical cyclone ends." Heiko Gernhauser Blog. www.wunderground.com/blog/JeffMasters/comment.html. May 16, 2007.
40. H. W. Whitaker (1985). "Citrus tree losses from 1983 and 1985 freezes in fourteen counties." *Proceedings of the Florida State Horticultural Society*: 98: 46–48.
41. K. A. Miller (1991). "Response of Florida citrus growers to the freezes of the 1980s." *Climate Research,* Vol. 1: 133–144, April 14.
42. Philip J. Klotzbach (2006). "Trends in global tropical cyclone activity over the past twenty years." *Geophysical Research Letters,* Vol. 33, LI0805, doi:10.1029/2006GL025881.
43. W. Briggs (2007). "ChangAmerican Meteorlogical Society. March 15, 2008, pp. 1387–1402, doi:10.1175/2007JCLI187.1.
44. C. Landsea, "How might global warming change hurricane intensity, frequency, and rainfall?" NOAA, Hurricane Research Division, Frequently Asked Questions. www.aoml.noaa.gov/hrd/tcfaq/G3.html
45. N. Scafetta (2010). "Empirical evidence for a celestial origin of the climate oscillations and its implications." J*ournal of Atmospheric and Solar-Terrestrial Physics,* doi:10.1016/j.jastp2010.04.015.
46. "Long-range solar forecast." (2006). NASA. science.nasa.gov/headlines/y2006/10may_longrange.htm
47. Ibid.
48. NASA, "Deep Solar Minimum." 2009, http://science.nasa.gov/science-news/science-at-nasa/200901apr_deepsolarminimum.
49. Ibid
50. "State of the Sun for 2008: ominously quiet –too quiet." 2/26/2009, http://www.sott.net/articles/

show/172348-State-of-the-Sun-for-2008-ominously-quiet-too-quiet.

51. "Cosmic rays hit space age high." (2009). NASA. science.nasa.gov/headlines/y2009/29sep_cosmicrays.htm
52. "Northern lights hit 100-year low point." (2010). AFP/file September 28. http://physorg.com/news204901293.html.
53. "A puzzling collapse of the earth's upper atmosphere." (2010). science.nasa.gov, July 15.
54. "Shrinking atmosphere layer linked to low levels of solar radiation." (2010). AGU release No. 10-28. www.agu.org, August 26.
55. Wikipedia, 8-23-2009. "Solar Variation.", http://en.wikipedia.org/wiki/Solar_variation.
56. B. Van Geel, et al. 1999, "The role of solar forcing upon climate change," *Quaternary Science Reviews* 18 (1999) 331-338.

Glossary

IN THE LAST 20 YEARS, we Americans and people around the globe have been subjected to every trick in the book in a concerted effort to convince us that mankind causes climate changes on the planet, and that the eternal, almighty Sun plays but a minor role. This grand hypocrisy has included a misuse of science and climate/weather terminology in a smoke-and-mirrors manner to convey such propaganda. The following definitions, with accompanying backgrounds, are offered in hopes of providing a fresh, realistic, and at times tongue-in-cheek look at how we define our environment, our climate, our world. It will also assist in serving as a baseline of understanding that will make the book more meaningful.

Anthropogenic Global Warming (AGW). A theory that postulates an ever-increasing warming of Earth's atmosphere and oceans so long as greenhouse gasses, in particular CO_2, produced by mankind's activity continue to rise.

Climate Change. The natural process of cyclical change in long-term weather patterns or climate on Earth, producing alternating warm and cold climate periods lasting decades to centuries. It is caused primarily by regular, repeating positional shifts in the Sun-Earth-Moon system and influenced by the gravitational effects of the other planets and variations in the Sun's activity levels.

After 2008, when the start of the next climate era of predominant cold weather started to become obvious, this term was used by AGW advocates as a replacement for "global warming." This was an obvious effort to mask the growing signs of the end of global warming and a climate shift to a cold era, and to lay the groundwork for a quiet, yet transparent retreat from the unsupportable AGW theory. This cover would allow AGW supporters to later claim, incredulously, that global cooling and global warming were both man-made and hence any kind of climate shift should be subject to legislative control.

Global Cooling. A process or state of cooling of the Earth's surface, oceans, and the lower troposphere, with associated climatic effects.

Beginning after 1945 but in particular during the early 1970s, the term became more specifically interpreted to mean the phenomena of dropping temperatures on Earth, the primary cause of which was erroneously thought to have been gasses from mankind's industrial activity, primarily sulfur dioxide (SO_2), producing a long-term "ice age" effect of planetary cooling.

Global Dimming. A normally short-term localized cooling effect produced by atmospheric sulfur dioxide and similar emissions often found near industrial sites or after volcanic eruptions.

These emissions or aerosols originate from both natural and man-made sources, which may have a real impact on climate, especially in isolated areas of the world, often associated with certain types of sulfur-rich volcanic eruptions. When present at the right altitude and concentrated in the upper atmosphere, the sulfur dioxide converts to sulfuric acid droplets, reflecting sunlight back into space and creating a cooling effect on the area of the Earth below the concentrations. This effect is most commonly witnessed after major volcanic eruptions and may produce global cooling lasting a few years.

Global Warming (GW). A process or state of progressively increased warming of Earth's surface, oceans, and the lower troposphere, with associated climatic effects.

Beginning in the late 1980s and up to 2008, the term "global warming" became more specifically interpreted to mean the phenomena of rising temperatures on Earth, the primary cause of which was erroneously thought to have been gasses from mankind's industrial activity, primarily carbon dioxide, producing a long-term "greenhouse effect" of continuous planetary heating. The term was misused and hence made synonymous with man-made global warming, also known as anthropogenic global warming (AGW). According to AGW supporters and reports of the United Nations Intergovernmental Panel on Climate Change (IPCC), mankind's greenhouse gasses were supposed to cause a continuous rise in Earth's temperature to the year 2100 and beyond. Predicted effects included melting of the world's glacial ice in Antarctica, Greenland, and high mountains, with resultant flooding of major coastal cities around the world. The term started to become politically risky to use after 2008 when a dramatic drop in Earth's temperatures was confirmed by monitoring stations and a new long-term downward trend in global temperatures appeared. This coincided with the formal announcement of the end of global warming at a news conference by the Space and Science Research Center (SSRC) in Orlando, Florida, on July 1, 2008.

As a result of further evidence provided by a host of scientists, it then became clear that the AGW theory was simply wrong. Its scientific basis became the subject of widespread, growing criticism. Between 2008 and 2009, the AGW theory was increasingly described by prominent scientists and leaders as a "scam," a "hoax," and a "fraud," especially after the revelation of manipulated data found during the "Climategate" scandal and significant errors in the science behind the IPCC reports. With the failure of global temperatures to rise above those of 1998, including through the warm year of 2010, a "no-growth" twelve-year history for Earth's temperatures had been established. The warming of the globe, or "global warming," had stopped.

Solar Hibernation. A period of time typically referred to by solar physicists as a "grand minimum." During these special, rare minimums, the Sun reduces its activity or output to a substantial degree for much longer — typically two to three normal solar cycles of 11 years each, or about 22 to 33 years. This hibernation results in significant global temperature reductions of historic proportions, with the potential to produce major ill effects worldwide, including social, political, agricultural, and economic disruption.

The term "solar hibernation" was introduced into the climate debate by John L. Casey, director of the Space and Science Research Center in Orlando, Florida, at a press release on September 22, 2008. It is associated with several key measurements of the Sun, most notably reduced sunspot creation, with sunspot peak cycle numbers in the 50 or lower range. This compares with past solar cycles at 150 or more sunspots at cycle peaks. Solar hibernations come along on average every 206 years and are among the most powerful solar minimums, resulting in major temperature reductions on Earth. The last solar hibernation was the Dalton Minimum between 1793 and 1830. The US government and major US media outlets were first notified of this next hibernation and its concurrent destructive effects by Casey in April of 2007. The predicted coldest year at the bottom of this next hibernation is estimated to be 2031, with potentially dangerous, record-setting cold either side of this low point. Many other scientists have confirmed Casey's prediction or have announced a similar forecast of a new cold climate era between 2010 and 2050.

The Theory of Relational Cycles of Solar Activity. A highly reliable theory that explains climate changes based upon solar activity and a family of relatively short cycles of the Sun that people can "relate" to during their lifetime. Of the many solar cycles, often lasting thousands of years, these Relational Cycles are those with durations around two centuries or less. It is also called the Relational Cycle theory or simply the RC theory.

Based upon an accidental discovery of the existence of a family of repeating cycles of the Sun's activity by John L. Casey in April 2007, the RC theory provides an accurate (>90 percent) explanation for major global climate fluctuations between warm and cold periods. According to the RC theory, future global climatic changes are generally predictable many decades in advance. The next climate change, correctly forecast in advance by Casey using the RC theory, has already started and is evidenced by dramatic changes in the Sun's reduced activity levels, the end of the past phase of continuous global warming, and a new trend of lower temperatures worldwide.